KNACK
MAKE IT EASY

NIGHT
SKY

KNACK

NIGHT
SKY

Decoding the Solar System, from Constellations to Black Holes

Nicholas Nigro

Photograph Research by Marilyn Zelinsky-Syarto
Illustrations by David Cole Wheeler

KNACK
MAKE IT EASY

Guilford, Connecticut
An imprint of Globe Pequot Press

Copyright © 2010 by Morris Book Publishing, LLC

Editorial Director: Cynthia Hughes
Editor: Katie Benoit
Project Editor: Tracee Williams
Cover Design: Paul Beatrice, Bret Kerr
Interior Design: Paul Beatrice
Layout: Kevin Mak
Photo Research by Marilyn Zelinsky-Syarto
Illustrations by David Cole Wheeler with the exception of those listed on pages 236–241
Front Cover Photos by (left to right): Courtesy of NASA; David Cole Wheeler; Courtesy of NASA; © Wikipedia Commons | photo by Senior Airman Joshua Strang
Back Cover Photo by: © Tatyana Chernyak | Dreamstime.com
Interior Photo credits on pages 236-241

Library of Congress Cataloging-in-Publication Data
Nigro, Nicholas J.
  Knack night sky : decoding the solar system, from constellations to black holes / Nicholas Nigro.
    p. cm.
  Includes index.

ISBN 978-1-59921-955-4
  1. Astronomy—Amateurs' manuals. 2. Solar system—Popular works. I.Title. II. Title: Night sky.
QB63.N62 2010
  520—dc22
                          2010011820

The following manufacturers/names appearing in *Knack Night Sky* are trademarks:
Celestron®
Coronado®
Meade®
Olympus®
Orion®
Post-it®
Thermos®

Printed in China

10 9 8 7 6 5 4 3 2 1

To the men and women of NASA for all that they do.

Acknowledgments

Foremost, I'd like to thank my agent, June Clark, for believing that I was up to this atypical task. Also, many thanks to my editor, Katie Benoit, for making this celestial adventure—storyboard and all—relatively pothole free. Mixed metaphors aside, special thanks, too, to photo researcher Marilyn Zelinsky-Syarto and illustrator David Cole Wheeler for bringing this intriguing subject matter to life with so many compelling visuals. Finally, plaudits to one and all at Globe Pequot who put this book together so quickly and capably.

CONTENTS

INTRODUCTION

The night skies have long fascinated the human species. Ancient civilizations worshipped the sun and other extraterrestrial bodies as gods. Sure, times have changed. Nowadays, life in the hustle and bustle of the twenty-first century frequently finds us overwhelmed by our daily routines and numerous responsibilities. We are often so busy going about our business that we fail to notice—let alone appreciate—the vastness and splendor of the solar system in which our planet resides. Yet, the night skies continually deliver spectacular shows, free of charge, with no repeat performances—and right outside our doors, too.

Look upon this book as a well-deserved and vivid ode to our solar system—a colorful picture window into the cosmic possibilities of stargazing: a hobby and passion that you need neither deep pockets nor prescribed locations to partake in. Its abiding mission is to assist you in experiencing the sheer magnitude, intricacies, and breathtaking beauty of our little niche in the colossal universe as well as in the remotest reaches of outer space.

Sunny Days, Starry Nights

Foremost, we cannot pay homage to our solar system without first saluting its prime mover: the sun. Let's face it:

We take for granted this shimmering orb in the sky, which comprises approximately 99 percent of our solar system's matter and choreographs its minutia. Indeed, the sun's gravitational muscle is a force to be reckoned with. It is the star of our solar system and, in reality, a yellow dwarf star in composition—one, interestingly enough, that is not wholly unique in a galaxy containing an estimated two hundred billion to four hundred billion stars. Nevertheless, this ultra-bright and exceptionally hot star, which has been in existence for more than four billion years, is what nurtures and sustains all life forms on Earth.

Prior to the sixteenth century, the accepted scientific wisdom held that humanity resides in an Earth-centered universe and that our planet is stationary. A Polish astronomer, Nicolaus Copernicus, begged to differ. He hypothesized that the focal point of our solar system is the sun—not Earth. Following on Copernicus's heels, Galileo Galilei validated his pioneering colleague's supposition. And key pieces to the night sky puzzle were soon added with contributions to the laws of physics from Johannes Kepler and Isaac Newton.

Considering the subject matter of this book, it's worth noting, too, that the aforementioned Galileo extensively employed a telescope to explore the width and breadth of the celestial heavens, thus establishing the yardstick for future examinations of the enigmatic night skies. And for a little perspective here: The telescopic capacity of Galileo's instrument was akin to today's very modest and reasonably priced telescopes—the kinds readily available and used by amateur astronomers.

Dynamic Night Skies

Along with this book's spectacular visual tour of our solar system and deep space, you will simultaneously be supplied with accessible tips and techniques to maximize the results of your stargazing endeavors. Not only will you learn about various astral accoutrements and equipment from star maps to telescopes, but also you will glean pertinent information on the differences in seasonal viewings of the night skies—in both hemispheres—as well as the precise dates of momentous celestial events in the offing.

Because the night sky is so infinite in its sprawl, amateur astronomers have actually made scientific discoveries of substance from their own backyards. Sometimes they are called "backyard astronomers." Even with thousands upon thousands of stargazers and scientists the world over perusing the celestial ether at any given moment, the night sky

is just too massive to fully cover. In fact, astronomy is the only branch of science where invaluable contributions from everyday people are commonplace.

Throughout the pages of this book, you will be visually engaged with images that, in a great many instances, you can check out for yourself in a night sky near you. And thanks to high-powered, sophisticated instruments like the Hubble space telescope and myriad space satellites that have snapped pictures from the far reaches of interplanetary and intergalactic space, you will also feast your eyes on more intimate and elaborate views of everything from our moon to constellations of stars to far-away galaxies.

The greatest and longest-running reality show is, in fact, the night sky. Sure, it's nice to have desktop wallpaper that features the Aurora Borealis, brilliant meteor showers, or solar eclipses, but it's even nicer to witness the real things in real time. This is precisely why more and more people, of all ages and in all corners of the world, are logging so many hours as dedicated amateur astronomers. The satisfaction gained in personally observing the cosmic players of the night sky is profound—and there is always a next level . . . and a next level after that. You will never run dry of observational fodder.

As youngsters, many of us peered into the night sky and saw the "man in the moon." Our youthful innocence and unfettered sense of awe and excitement had us checking out the Big Dipper and Leo the Lion, too. But here's something interesting to mull over: Amateur astronomers long removed from their childhoods will tell you that the thrills of combing the night skies have not been cast asunder by world-weary adulthoods. Stargazing, it seems, has a knack for simultaneously resurrecting childhood enthusiasm and marrying it with a more in-depth understanding and respect for all the goings-on in the busy night sky. It's a potent one-two punch that makes this hobby a true original.

Stargazing is exciting, enlightening, and thought-provoking. The more you encounter in the night skies, the more you wonder. The more you ponder the dynamism of our solar system, the more you wonder. The more you reflect on our unique place in the cosmos, the more you wonder. The night skies are full of wonder—that's the long and short of it. The hundreds of pictures in this book covering scores of regions and objects in our solar system, as well as in farther reaches of the universe, will captivate you and make you think. When you contemplate the enormity of what you are surveying in outer space, planet Earth emerges as a mere speck. The sun is just one of billions of stars in our galaxy, the Milky Way; the Milky Way, just one galaxy among billions beyond its borders. Amateur astronomers can see things that are millions of light-years away. In other words: Looking into the celestial beyond literally furnishes you glimpses into what once was.

The Sky's the Limit

There are no hard and fast rules for stargazing. You don't have to boldly go where no astronomer has gone before to make night-sky watching a truly rewarding experience. You don't need a scientific mind or specialized college degree to inspect the night sky. You can study anything you want on the celestial sphere from virtually anywhere. You can observe the stars up above with only your two eyes, or you can use enhancing gear to get a closer look. If you so desire, there are amateur astronomy groups to join, or you can stargaze all alone, with friends, or with members of your family.

Use this book as a handy and informative tour guide of the night skies, inviting you to further explore on your own, or in the company of like-minded souls, our solar system and the mystifying universe at large. After you have sampled from the night sky's eclectic menu, you will want to go back for seconds and thirds. You will never tire of locating individual stars, planets, constellations of stars, and galaxies a long, long way from home.

You can probe entities as nearby as the moon or as far away as deep-sky objects multiple millions of light-years away. You can use star maps, practice the art of star hopping, bone up on astrophysical principles, or just step out into your backyard right now and cast your eyes into an animated and unfathomable interstellar existence far removed from the cozy confines of planet Earth.

SUN, MOON & STARS
These bodies are key portals to understanding how our solar system works

In today's day and age, the sun, moon, and stars are more often seen as the stuff of commercial and media brands than celestial objects of colossal importance. That is, associated in many people's minds with movies, media, and merchandise ranging far and wide from software to footwear. Yet, when we pause and reflect on what is literally transpiring right outside our windows and beyond the boundaries of our planet's exosphere, it's downright mind-boggling. How most life forms on Earth are supported is a fascinatingly layered business.

Our sun is a star—the closest one, by far, to Earth. Yet, it is still an average 93 million miles (150 million kilometers) away. The moon, our planet's only natural satellite, is—relatively

The Sun: Earth's Lucky Star

- Imagine a sun one thousand times brighter than ours! Although our sun's diameter of 864,000 miles (1,390,000 kilometers) is immense, the biggest stars in the universe are one thousand times its size.

- Our sun, which consists of myriad gases, is approximately 4.6 billion years old. It has, hopefully, another four to five billion years of good living left.

- The sun's radius—the distance from its super-hot nucleus to its surface, known as the "photosphere"—is 432,000 miles (696,000 kilometers).

Earth's Moon

- The only celestial body that human beings have personally visited is the moon.

- The word *moon* is derivative of the Latin word *mensis*, meaning "month."

- There are two sides to the moon. We see only one side of it: its "near side," never its "far side."

- The moon's contrasting light and dark hues give it a distinct countenance, which human societies long ago dubbed the "man in the moon."

speaking—a hop, skip, and jump away at an average of 238,000 miles (383,000 kilometers) away. And the countless stars we see twinkling up above are light-years away from Earth. Interestingly, these benign-appearing outer-space bodies, which regularly cast gossamer blankets of glitter across the night skies, sport compositions similar to our sun's. In other words: Stars are hardly warm and fuzzy objects. But fortunately, we don't really have to worry about their blistering interiors and super-high temperatures, which generate nuclear fusion. We can appreciate their brilliance from a proper distance. Surveying them with our own two eyes, a pair of binoculars, or a telescope will suffice.

We can simultaneously explore the celestial beyond and contemplate the reality that our sun, like all stars, has a finite lifespan, which means that someday it will cease to be the guiding force and master of our solar system. No need to worry, though: That dark day is billions of years down the road. For right now, we can place the sun, moon, and stars on fitting pedestals. We can respect their roles and places in our solar system, as well as in the imposing and frequently bewildering universe at large.

Star-filled Night Sky

- Although it's only an educated guess, one estimate puts the number of stars in the universe at 70 sextillion, or 70,000 million.

- Stars appear to twinkle thanks to refracted light. Earth's layers of atmosphere twist light like a pretzel.

- Stars actually enlarge with age and often appear brightest before shuffling off the celestial coil for good.

- Converting hydrogen into helium throughout their enduring lives, stars, like our sun, are veritable nuclear reactors in the extraterrestrial ether.

Goodbye Sun, Hello Moon

- Just like the setting sun, when near the horizon, the moon appears larger than normal to our probing eyes. It is, however, slightly smaller and farther away from Earth at that point than when it is high in the sky.

- Although the moon shines brightly in the night reflecting sunlight, it is, as solar system bodies go, a poor reflector of light.

- The Soviet Union reached the moon first with a spacecraft, but sans any humans on board.

PLANETS

The planets in our solar system are a diverse, dramatic, and mysterious lot

A planet is a celestial object orbiting a star. Our planet Earth, for instance, orbits the sun, which is a yellow dwarf star, and so do seven other planets: Mercury, Venus, Mars, Jupiter, Saturn, Uranus, and Neptune. Pluto, which was once classified a planet, has recently been downgraded to "dwarf planet."

Since recorded time, these particular space bodies have captivated humankind. In fact, the word *planet* is a derivative of the Greek word meaning "wanderer." Ancient civilizations viewed planets as either divine or, at the very least, agents of the divine. And although we may not see them as such today, the various planets that inhabit our solar system still intrigue us.

Eight Diverse Planets

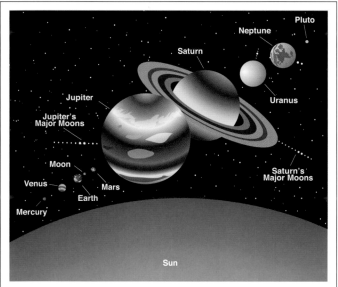

- Jupiter flaunts more than double the matter of the seven other planets combined.

- Mercury, the smallest planet in our solar system, is .38 Earth's size.

- At .95 Earth's diameter, Venus is practically our parent planet's twin in diameter.

- The inner planets—Mercury, Venus, Earth, and Mars—are significantly smaller than the outer planets, the so-called gas giants of Jupiter, Saturn, Uranus, and Neptune.

Moonstruck

- When the moon is up, it is the brightest object in the night sky.

- Venus is the brightest planet. It is often visible in the early mornings or early evenings depending on its orbital position.

- Venus is always fairly low in the sky and never visible throughout an entire night. Jupiter, on the other hand, is discernable at various points in the sky all night long.

- Jupiter is the next brightest night-sky object after the moon and Venus.

The youngest of school kids know that the planet Saturn has rather remarkable rings around it. And Mars's mysterious "face" on its rocky surface prompts speculation as to whether there was once life on that planet, or whether there is life on any of the others. When we peer into the night sky and spy a nontwinkling celestial object, we know we are witnessing a planet because stars twinkle and planets don't. The planet Venus, for instance, has been deemed both the "Morning Star" and the "Evening Star" because of its distinctive and bright appearance in the sky.

The inner planets—those closest to the sun—are predominantly made of rock, the byproduct of unremitting solar wind tumbling off the sun and careening through interplanetary space. The outer planets, farther from the sun's solar battering, are entirely different in makeup and composed of mostly gases.

The planets one and all nonetheless reveal a great deal about the inner workings of our solar system. They furnish us, too, with intriguing visuals that encourage us to want to see more but also to know more about their present existence, past life, and how they came to be.

Planets and the Moon

- The moon is approximately one-quarter of Earth's size.

- The smallest planet, Mercury, which is also closest to the sun, is only slightly larger in dimensions than the moon. The moon is roughly one-half the diameter of Mars.

- Mercury's surface is similar to the moon's cratered and rocky terrain. Like the moon, it is a poor reflector of sunlight.

- Although the moon is not classified a planet, it is nonetheless comparable in composition with the rock-hard terrestrial planets.

Saturn's Incomparable Rings

- Courtesy of its conspicuous and colorful rings, Saturn is perhaps the best-recognized planet. There are, in fact, thousands of rings orbiting Saturn.

- Jupiter, Uranus, and Neptune also have rings surrounding them. However, they aren't nearly as large and spectacular as Saturn's.

- Each one of Saturn's rings orbits the planet at a different speed.

- Rings on planets are composed of pieces of ice and rock derivative of comets, moons, asteroids, and myriad celestial objects.

NATURAL SATELLITES

Planets orbit the sun, and natural satellites orbit planets and other smaller celestial bodies

Although there are eight planets, including Earth, orbiting the sun, there is another considerable class of orbiters in our solar system. To differentiate this celestial fraternity from the human-designed and empowered contraptions known as "satellites," which prowl the corridors of outer space and snap all kinds of photographs and take all kinds of readings, these bodies are dubbed "natural satellites."

Natural satellites are celestial objects that orbit a planet or smaller space body like an asteroid. The term *natural satellite* is frequently used interchangeably with *moon*. In fact, the Earth's moon has the distinction of being among the largest natural satellites in relation to its parent planet's dimensions.

Celestial Partners

- The moon, Earth's only natural satellite, has been the subject of observation and conversation since prehistoric times.

- The most conspicuous consequence of the Earth-moon marriage is the tides. Tidal ebbs and flows are the result of Earth's gravitational tug-of-war with the moon.

- The moon orbits Earth synchronously, always keeping one hemisphere facing its parent planet.

- The moon's gravitational sway is more persuasive on the side of Earth closest to it.

Galilean Moon Io

- With its active volcanic surface replete with fresh lava flows and lava lakes, Io is one of the most intriguing moons in our entire solar system.

- Io has mountains as high as 52,000 feet (16 kilometers).

- Io is the nearest of the so-called Galilean moons to its parent Jupiter.

- The Galilean moons of Jupiter—the largest of the planet's numerous natural satellites—received their moniker from the man who discovered them: pioneering astronomer Galileo.

The moon is approximately one-quarter the size of Earth. Only Pluto's moon Charon, which is half its parent's size, is proportionally larger.

Nevertheless, as is so often the case in the study of the wider universe, all things are relative. The biggest known moon in our solar system is Jupiter's Ganymede. Both Ganymede and Saturn's Titan moon are actually greater in size than the planet Mercury. And Mercury and Venus, the two nearest planets to the sun, have no moons shadowing them, whereas Mars counts two small moons in its orbit.

ZOOM

From our vantage point here on Earth, we can see only one side of the moon—near the side. The so-called far side of our planet's only natural satellite is never visible to us. It's a hidden hemisphere perpetually turned away from Earth. A veritable monolith, the far side of the moon maintains an almost uniformly cratered surface.

Mars's Smaller Moon

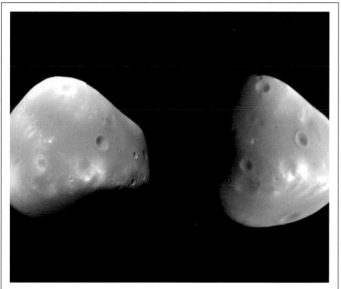

- Many scientists surmise that both Deimos and Mars's nearer, larger, and more studied moon, Phobos, are former asteroids long ago captured by the planet's gravity.

- Deimos is among the smallest known moons in our solar system.

- *Deimos* is a name out of Greek mythology. It literally translates as "panic." Many moons and celestial objects sport names from ancient works of literature.

- Among the four inner planets, Mars and its two moons take first prize.

Pluto's Moon or Double Planet

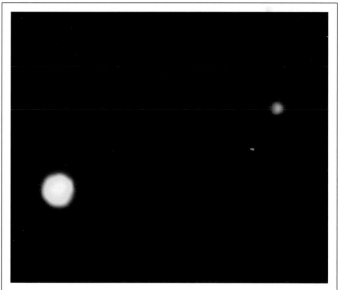

- Pluto, once considered a planet, was recently demoted by the International Astronomical Union (IAU) to "dwarf planet" status. It nonetheless has three natural satellites: Charon, Hydra, and Nix.

- Some scientists speculate that Charon is not a moon but rather part of a double planet.

- Planets aren't the only celestial bodies with moons. Some asteroids have natural satellites.

- Charon's diameter of 750 miles (1,207 kilometers) is half that of Pluto.

COMETS & METEORS

Comets and meteors supply stargazers with colorful clues to the makeup of outer space

Witnessing a comet's appearance in the night sky is considered a stargazing moment to remember. A meteor shower is a natural light show worth the price of admission. There is nothing quite like seeing with your own two eyes a comet in motion or a dazzlingly unpredictable meteor shower.

Comets consist of a hodgepodge of space debris, including myriad gases, ice, rocks, and dust. And they don't hang around for long periods of time, either.

Comets can be seen in the night sky as their paths journey nearest to the sun and their icy-cold nuclei begin to melt, generating gaseous auras called "comas" and ionized and sullied tails trailing in their wake, which reflect sunlight back to Earth.

Comet in its Orbit

- Comets are always unevenly shaped. Even naked-eye observations can detect their highly irregular contours.

- Most comets travel in pronounced elliptical orbits around the sun, traveling very near and then very far away.

- Some comets venture too near the sun and meet their cosmic makers. They are known as "sungrazers."

- In ancient times, comets were considered harbingers of bad things to come.

Comet: An Intimate Look

- When comets approach the sun, luminous tails consisting of gas and dust are the byproduct.

- Comets' celebrated tails always face away from the sun. The powerful solar wind incessantly streaming off the sun's exceptionally hot surface sees to that.

- The tails of comets shrink and then fade away altogether as they move away from the sun.

- Comets are visible to us when they are near the sun but practically indiscernible during the remainder of their orbits.

Meteors, on the other hand, often appear in the night sky as wayward stars knocked off their fixed celestial perches and are frequently referred to as "shooting" or "falling" stars. They are, however, pieces of rocky or metallic space matter—far removed from stars and their blistering hot interiors. When orbiting meteoroids permeate the Earth's atmosphere, they officially become meteors as they burn up. Occasionally meteors don't completely fragment in the atmosphere's vise and crash into our planet's surface.

ZOOM

Although the exact origins of comets are still only scientific conjecture, their overall compositions suggest a correlation, in many instances, with the formation of the outer planets—Jupiter, Saturn, Uranus, and Neptune—which are believed to be approximately 4.6 billion years old.

Meteoroids Generating Meteor Showers

- As you read these words, meteoroids are crashing through Earth's atmosphere.

- Most meteoroids are no bigger than a pebble and some are even as small as a grain of sand.

- Meteoroids orbit the sun, but because of their varying sizes, their orbiting patterns radically fluctuate.

- When meteoroids make it down to Earth's surface, they are called "meteorites."

Phenomenon in the Night Sky

- Meteoroids burn up in Earth's atmosphere and prompt streaks of light called "meteors."

- Meteors can be seen during all four seasons of the year, although there are specific times when annual meteor showers occur.

- The word *meteor* is derived from the Greek word *meteoron*, which translates as "phenomenon in the sky."

- Meteor showers occur when our planet's orbit intersects with a comet's or asteroid's trail of debris.

GALAXIES

Billions of galaxies throughout the observable universe provide a window into its origins

We not only live in cities, countries, and on a planet called "Earth" but also amidst a sprawling galaxy. More specifically, in the Orion arm of a galaxy known as the "Milky Way."

Galaxies are distinctive groupings of stars and interstellar material all bound together by gravity. And although we can see the Milky Way in the night sky, which often appears as an elongated "milky" arch of light, we cannot observe it in its entirety. Why? Because Earth is a planet inside of a solar system, which is also inside of an expansive galaxy's disk of stars and dust. In other words, what we are witnessing when surveying the Milky Way is only a portion of our parent galaxy. We are, in essence, on the inside looking out.

Pinwheel Spiral Galaxy

- Approximately 77 percent of the observed galaxies in the universe are classified as spiral galaxies.

- Courtesy of its considerable and prolific star-forming areas, the Pinwheel galaxy has been christened a "grand-design" spiral galaxy.

- The Pinwheel galaxy is 27 million light-years away but nonetheless measures 170,000 light-years across.

- Amateur astronomers can see the Pinwheel galaxy with a small telescope. But, because of its distance, it appears as only a dim blotch in the night sky.

Barred Spiral Galaxy NGC 1300

- It is estimated that close to half the observed spiral galaxies sport elongated bars across their midsections.

- These bars are composed of stars whose orbiting patterns, it is surmised, are tightly choreographed by the galaxies' dynamic galactic centers.

- The bar structures detected in many spiral galaxies are not considered permanent features. It is believed they will disappear with time.

- The Milky Way galaxy features a small bar of bright stars close to its nucleus.

The Andromeda galaxy is our closest galactic neighbor and can be seen with the naked eye in Northern Hemisphere night skies. It is approximately 2 million light-years away. In the Southern Hemisphere, the irregular galaxies known as the "Large Magellanic Cloud" and "Small Magellanic Cloud," at some 160,000 to 180,000 light-years away, can also be seen without the aid of telescopes.

Galaxies throughout our universe number in the billions. The most distant one—recently discovered by the Hubble space telescope—is an estimated 13 billion light-years away, which is not far removed from the age that scientists place the universe. The astronomical community is especially intrigued by galaxies because of the sum of their parts, which includes baffling black holes in their galactic centers. For some perspective, the black holes in many galaxies are presumed to be up to one billion times as massive as our sun. And the sun encompasses close to 99 percent of our solar system's total matter. When you do the arithmetic, it all adds up to a lot of space in outer space—almost beyond our comprehension.

Elliptical Galaxy NGC 1132

- Elliptical galaxies, which are ellipsoidal in shape, have no spiral arms to speak of.

- Most elliptical galaxies accommodate fewer stars than spiral galaxies and tend to reside in clusters of galaxies.

- Whereas spiral galaxies showcase bluish hues in their spiral arms, elliptical galaxies often appear yellow-reddish in color.

- Yellow-reddish hues typically indicate the presence of older stars; bluer colors, younger stars.

Irregular Galaxies

- M82 is the only irregular galaxy that appears in the renowned catalog of celestial observations known as "Messier's objects."

- From scientists' perspectives, irregular galaxies exhibit no defined shapes—that is, no spiral or oval contours.

- Only 3 percent of observed galaxies do not make the grade as either spiral or elliptical. A distinct minority of galaxies is classified as irregular.

- The Large Magellanic Cloud qualifies as an irregular galaxy.

PLACE IN THE COSMOS

Earth is a speck in a spectacularly expansive—and possibly expanding—universe

We inhabit the third rock from the sun. Mercury and Venus are first and second. And Mars follows on the Earth's heels as the fourth rock. Among these four so-called inner planets—also known as the "terrestrials" because of their rocky and metallic compositions—the Earth reigns supreme in size.

However, the outer planets—dubbed the "gas giants"—of Jupiter, Saturn, Uranus, and Neptune paint an entirely different portrait in both mass and makeup. Jupiter alone constitutes more than two and one-half times the mass of all the other planets in our solar system combined. And our solar system is a mere shard in the Milky Way galaxy, estimated at 100,000 light-years in diameter. Simply understood, it would

Moon Perspective of Earth

- The *Apollo 8* astronauts were the first humans in a lunar orbit. They snapped pictures of Earth from the unique perspective of a lunar horizon.

- The *Apollo 11* mission landed on the moon on July 20, 1969. The only outer-space entity, other than the moon, which astronauts could conceivably visit someday is the planet Mars. However, before this occurs, the monumental quandary of intense space radiation must be addressed.

- A space expedition to Mars could take more than a year.

Solar System Troupe

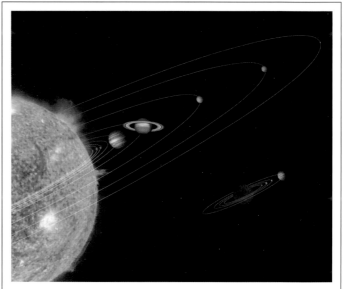

- Our solar system is the region in space dominated by the sun's gravitational muscle. All planets and celestial objects (including comets, asteroids, and meteoroids) orbit the sun.

- At various boundary points, the sun's impact is no longer felt. It is here where the outer limits of our solar system exist.

- Our solar system is an infinitesimal sliver in the universe at large.

- Other solar systems with stars and orbiting planets, long suspected, are now documented reality.

take—traveling at the speed of light—roughly 100,000 years to trek across the width and breadth of the Milky Way.

To add to this rather rambling portrait of the stellar beyond, space telescopes have uncovered billions of galaxies in the universe, including some more light-years away than planet Earth's determined age. In other words, galaxies multiple billions of light-years away, revealing the existence of celestial bodies long before there was a planet Earth, can be seen and photographed. And this reality snapshot alone is enough to turn the hobby of stargazing up a notch.

Our universe is, in fact, so massive that it cannot ever be accurately measured. Many scientists believe that it is actually expanding in size and that it may not be the only one of its kind. So, our unique place in the cosmos—as the only demonstrable life—is on wobbly ground. Can we really and truly believe that the only life forms in an immeasurable universe of megaproportions are on one tiny planet called "Earth"? And if there are other universes yet uncovered, surely the odds of some kinds of life elsewhere increase dramatically.

Starwatching in Antiquity

- Astronomy is among the oldest of natural sciences, dating back to antiquity.

- Before the invention of the telescope, naked-eye observations of the night sky were the only game in town, although various sighting tube devices were employed.

- Ancient astronomers early on distinguished between the visible planets and stars.

- Ancient Greek astronomers, including Hipparchus of Nicea in the second century B.C. and Ptolemy of Alexandria in the second century A.D., gave birth to mathematical astronomy.

Today's Sky Watchers

- We can survey the Sun in the daylight hours and the vastness of outer space after dark.

- We can see things millions of light-years away with our naked eyes, binoculars, and telescopes.

- The universe is mobile.

Earth is revolving around the sun; the sun is orbiting around its galaxy's center.

- The Hubble space telescope has its scopes on the outer reaches of the universe, snapping pictures of space bodies older than the universe itself.

OPTIMAL WEATHER CONDITIONS

To experience all that the night sky has to offer, the right weather matters

Maximizing your nighttime starwatching requires some serious cooperation from Mother Nature. Obviously, the clearer the night skies, the more you will able to see. But optimal weather conditions involve more than the absence of storm clouds interfering with your celestial observations. A consistent breeze will hinder your viewing pleasure. Even during a clear and

cloudless night, winds can stir up the atmosphere and detract from your experience. You could be blissfully unaware of the negative ramifications of wind, too. In any event, consider a still night the best kind of night to cast your eyes into the skies.

And because many amateur astronomy outings occur during the summertime, when the night skies put on some of their

Clear Night Sky and Full Moon

- Before planning stargazing events, it's always a prudent idea to pay special heed to weather forecasts.

- The optimum starwatching conditions embody not only cloudless skies but also low relative humidity levels.

- The American Southwest's Sonoran Desert, Bryce Canyon in Utah, Grand Canyon in Arizona, and Yosemite and Death Valley in California are considered night-sky country. Utah's Natural Bridges National Monument has actually been christened an "International Dark Sky Park."

Benefits of Limited Moonlight

- A thin crescent moon or—better yet—no moon at all is best for starwatching.

- A sky-high moon bathes the nighttime skies with light, reducing stargazing sightings to only the brightest stars and planets.

- Unless the moon is your stargazing prey, its new moon phase is the ideal time to peer into the celestial ether.

- Most astronomy clubs host their star parties between the moon's last quarter and new moon phases.

best shows, there is often a matter of relative humidity to contend with. The bottom line is that less is better when it comes to humidity. Excessive moisture in the air cannot help but cloud the best possible picture window into the night sky.

It also should be noted that unless the moon is your stargazing target, the less moonlight there is, the better. You don't need the night sky illuminated for you by a celestial object only 238,000 miles away. Moonlight diminishes the all-important darkness that supplies you with the optimal portal into the vast frontier that is outer space.

Perfect Stargazing Blueprint

- Chile's Atacama Desert is the highest desert on the planet and a stargazing mecca.

- Although it's not easy to get to, the Atacama Desert marries high altitude with dry air and no vestiges of light pollution: the perfect blueprint for seeing all that

the night sky has to offer.

- Humidity is a frequently overlooked adversary of starwatching.

- On hot and muggy summer evenings, it's better to be cooling it in air-conditioned indoors than stargazing in the great outdoors.

Wind and Cloud Cover

- Windy nights are not recommended times to check out what's playing on the celestial sphere.

- Wind churns up the atmosphere and precludes you from seeing all that you would otherwise see on a windless evening.

- Excessive wind also wreaks havoc on telescope users, who, above all else, benefit from climate tranquility.

- Although beaches are often atmospheric stargazing locales, they are also prone to be windy places. Moving away from the water's edge is sometimes a smart move.

FAVORABLE LOCATION

The real estate mantra "location, location, location" is germane to your stargazing endeavors

The beauty of the stargazing hobby is its simplicity. You can enjoy this hobby from virtually anywhere, although obviously not all locales are created equal. A general rule of thumb is this: Rural is better than urban; clean air is better than foul air.

For starters it's a prudent idea to seek out a safe and stable location to survey the night skies. That is, rooftops and other high elevations may furnish you with a fine view, but they are sometimes dangerous places to be when moving back and forth. Furthermore, you should be completely comfortable when combing the celestial ether, and this includes wearing proper dress for the season of the year and temperature. And depending on the duration of your stargazing escapades,

Safety First

- National parks are popular places for amateur astronomers to gather because they furnish premium views into the night skies in safe, spacious areas.

- Rooftops are stargazing hotspots, but unless they are completely secure, it's a good idea to set up shop on lower, more stable grounds.

- Before landing in a remote location in the dark of night, do your homework.

- If feasible, stargaze in a known park or open field as far away from artificial lighting as physically possible.

Suitable Attire

- If you are stargazing in the cold climes of wintertime, be especially certain that you are suitably attired.

- Starwatching involves a fair share of immobility. So dress with that physical level of activity—or inactivity, as it were—in mind.

- Stargazing is a nighttime event. It can be downright chilly sometimes.

- Although desert locations are popular stargazing locales, amateur astronomers are frequently unprepared for the cool temperatures that accompany desert nights.

appropriate supplies may be in order—food, for instance, and a Thermos of hot coffee on a chilly night.

The proper location also has an unobstructed view of the night sky—or as unobstructed a view as possible. A panoramic view of the entire sky is the ultimate. However, such a sweeping vista is not always achievable. Look at it this way: The more sky you can physically observe, the more celestial regions, objects, and happenings will be at your disposal.

YELLOW LIGHT

If you are interested in the nightlife, then a big city that never sleeps is what you are after. For stargazing, however, the bright lights of big cities are the last things you want. Artificial lighting from the human populace either at work or play, which casts its luminescence into the night sky, dramatically reduces what you can observe up above.

Unobstructed Views

- If you can glance north and south then east and west and spy the celestial sphere in its entirety, you are in a prime stargazing locale.

- Don't sit under a tree while stargazing. Open fields or elevated locations with no obstructions are the best locales for supplying you

with a panoramic view of the night sky.

- Nearby buildings or structures of any kind interfere with a full view of the celestial beyond.

- Higher altitude locations naturally deliver the best views of the night sky.

Artificial Lighting

- Bright lights, big city: bad for stargazing.

- Locations in close proximity to large cities are likewise not premier locations for starwatching.

- It isn't just cities with excessive lighting that make night-sky viewing less than

optimal, but also merely being near streetlights is enough to diminish the experience.

- Car lights also play a role in what you can and cannot see in the night sky. It's better to be as far away from automobiles as possible.

BASIC GEAR

The two most readily available tools for surveying the night sky are the naked eyes

Some people view astronomy as a cost-prohibitive hobby—an overly expensive pastime that will break the bank. They associate peering into the night sky with pricey equipment like high-powered telescopes, detailed star charts, and all kinds of astrophotography accoutrements. But the reality is that you can see an awful lot with just the naked eye.

In addition, ordinary fold-up outdoor chairs, with reclining backs, are basic gear that will allow you to comfortably adjust your head view as needed. Really, for stargazing purposes, a costly ergonomic chair is not required.

Indeed, on a clear night in a location away from city lighting and pollutants, your two eyes—with no magnifying

Naked Eyes

- Before purchasing a pair of binoculars or a telescope, many professional astronomers recommend naked-eye observation of the night sky.

- Without enhancing aids, learn where key markers in the night sky, like the North Star and the Big Dipper, are.

- Even some deep-sky objects can be seen with the naked eye under optimal conditions.

- The unaided eye is unencumbered and permits you unlimited flexibility in perusing the celestial sphere east to west and north to south.

Binoculars

- Before purchasing a telescope, comb the night skies with binoculars.

- Binoculars are easier to use than telescopes and thus better suited for fledgling amateurs in the hobby.

- Starwatchers frequently employ binoculars with 7 x 50 or 10 x 50 magnification.

- The wider the field of view on the binoculars, the more celestial light the instrument will gather, which is what you want.

devices—can decipher constellations of stars, planets, star clusters, and other deep-sky objects. In the Northern Hemisphere, it's possible—under the right conditions—to see our neighboring galaxy, the Andromeda, without binoculars or a telescope. And then there's the Milky Way, which is readily seen with the naked eye.

Of course, if you want to kick your stargazing adventures up a notch, a basic pair of binoculars will do nicely in enhancing your night-sky observations, permitting you to see fainter deep-sky objects.

YELLOW LIGHT

It's highly recommended that you stargaze in the night with a red flashlight. This kind of light does no harm to night vision. On the other hand, most white lighting after dark, even the smallest flash of a lighter or a match, sets your night vision back. Your pupils need to be fully dilated to see best in the dark. They will adjust in total darkness—after approximately thirty to forty-five minutes—to give you the sharpest view of the night.

Red Flashlight

- Stargazers should carry a red LED flashlight, which is expressly designed with the astronomer in mind.

- If you can't access a red flashlight, cover an ordinary flashlight with red cellophane.

- A no-battery flashlight, which is not as bright, is also preferable to a conventional battery-operated flashlight.

- Remember: Flashing any kind of white lighting at a stargazing party will make you persona non grata in a New York minute.

Reclining Chair

- Foremost, to maximize your starwatching, you should be as relaxed as possible.

- If you are planning an extensive stargazing adventure, a comfortable chair is in order. Standing for hours upon hours can get rather tedious and stress your neck and back.

- If you are using binoculars, a reclining back on a chair can affix your head position to your liking.

- There are actually chairs, binocular chairs, specifically designed with the stargazer in mind.

TELESCOPES

These instruments take stargazing to a more in-depth and intimate level

The telescope is practically synonymous with the science of astronomy. Although a telescope is not required to survey the night skies, the right one could enable you to see things clearer and with more contrast as well as see things that you wouldn't otherwise see with the naked eye.

There are countless styles of telescopes that range far and wide in size, appearance, and price. Nonetheless, all telescopes are classified as one of three kinds: refractors, reflectors, or catadioptrics.

Simply understood, refractor telescopes employ lenses that refract light. That is, they "twist" light as it accumulates at their considerable heads—that is, the objective lenses. Refractor

Refractor Telescope

- Small refractor telescopes are often recommended for beginners.

- A good starter telescope often sports 2.4-inch (60 millimeter) or 3-inch (80 to 90 millimeters) refractors, which will furnish you with ample observational grist.

- Although binoculars are fine for viewing the night sky, they do not supply you with any detail. Telescopes afford intricate looks.

- Keep in mind that you are always observing space bodies through Earth's layers of atmosphere, which distort things.

Reflector Telescope

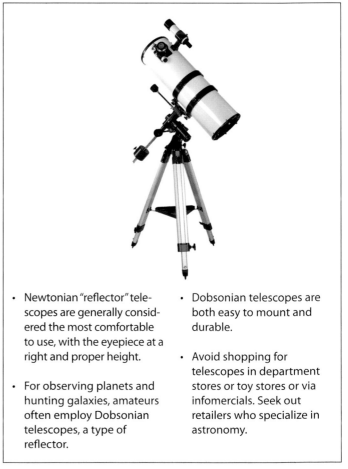

- Newtonian "reflector" telescopes are generally considered the most comfortable to use, with the eyepiece at a right and proper height.

- For observing planets and hunting galaxies, amateurs often employ Dobsonian telescopes, a type of reflector.

- Dobsonian telescopes are both easy to mount and durable.

- Avoid shopping for telescopes in department stores or toy stores or via infomercials. Seek out retailers who specialize in astronomy.

telescopes are rather simply designed, weighty, with minimal maintenance requirements. They are thus expensive. Refractors are fine telescopes for general night-sky viewings but are not the best telescopes for tracking down deep-sky objects.

Also known as "Newtonian telescopes," reflector telescopes, on the other hand, use mirrors to accumulate light. These telescopes are renowned for supplying their users with panoramic viewing fields, significantly wider than their refractor competitors. Reflectors are excellent for locating dim deep-sky objects, including distant galaxies and nebulae. Because

they use mirrors instead of more costly lenses, Newtonian telescopes are typically the cheapest on the market. There are many varieties of these telescopes with widely varying capacities. The Hubble space telescope is actually a sophisticated version of a Newtonian telescope.

Finally, there are catadioptric telescopes, which combine the best of both the refractors and reflectors while largely torpedoing their negatives. Because it is not central casting's image of what a telescope should look like, a catadioptric telescope might appear a tad strange to you.

Catadioptric Telescope

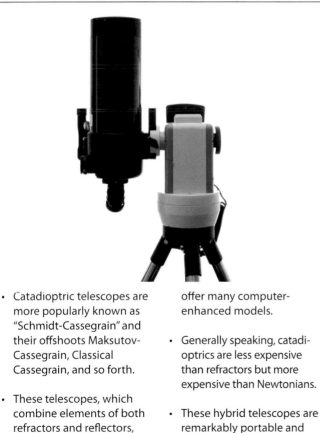

- Catadioptric telescopes are more popularly known as "Schmidt-Cassegrain" and their offshoots Maksutov-Cassegrain, Classical Cassegrain, and so forth.

- These telescopes, which combine elements of both refractors and reflectors,

offer many computer-enhanced models.

- Generally speaking, catadi-optrics are less expensive than refractors but more expensive than Newtonians.

- These hybrid telescopes are remarkably portable and offer myriad accessories.

Telescope at Work

- There are three key aspects to assess when purchasing a telescope: light-gathering capacity, quality of the optics, and ease of mounting.

- Observing the moon with a telescope can furnish you with incredible detail.

- Stars appear brighter in a telescope but not any bigger. They are just too far away.

- Some popular manufacturers of telescopes include Meade, Celestron, and Orion.

ADVANCED STARGAZING
As you become more experienced in the hobby, you can kick it up a notch

While exploring the night skies, there are no ironclad rules as to how you approach the hobby or what you should be checking out in the celestial ether. Nevertheless, for many stargazing practitioners, fully appreciating the wonders of outer space requires a certain amount of knowledge of what things are, where they are, and how they impact their immediate environs. To advance your stargazing education, you can begin with simple star maps, also known as "sky maps." A star map identifies the positions of constellations and the stars therein and enables you to pinpoint exactly where these outer-space areas and bodies can be found.

Star hopping is another technique stargazers employ to

Star Map

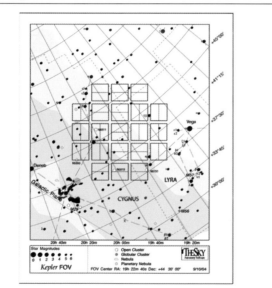

- To get the utmost out of star map use, it's important to know your latitude and longitude on Earth.

- On-demand star map Internet sites will print out a chart based on your exact observational location and specific date.

- Star maps enable you to piece together the sprawling night sky into stellar neighborhoods.

- After you know where things are—and where exactly to look—your stargazing time wasted is considerably reduced.

Star Hopping

- Polaris, the North Star, is a popular star-hopping reference point in the Northern Hemisphere.

- The art of star hopping, with an accompanying star map, is a learning process.

- Star hopping has been analogized to learning to ride a bicycle. That is, after the technique is mastered, you will not ever forget how to effortlessly side-step from key celestial objects or areas in space to other objects and areas.

- All you need to star hop are eyes, arms, and fingers.

assist them in deciphering various regions of the night sky. This technique amounts to locating the brightest and most familiar objects in space, knowing exactly what they are, and—here's the kicker—what is around them in every direction. Star hopping enables you to use celestial bodies as guideposts to forage farther afield in the surrounding skies. This detection method often goes along with star maps. An example of star hopping would be locating the well-known asterism the Big Dipper in the night sky and then moving from there to identify nearby stars, constellations of stars, and other objects of interest.

Advanced stargazing is also enhanced with some insight into the laws of physics as they apply to astronomy. The night sky is a whole lot more than a captivating picture show. There's a great deal occurring in outer space with the forces of gravity, interacting gases, nuclear reactions, splitting atoms, and so forth at work. Even the most fundamental knowledge of astrophysical principles will take your perusal of the night skies to a more in-depth level.

The Law of Gravity

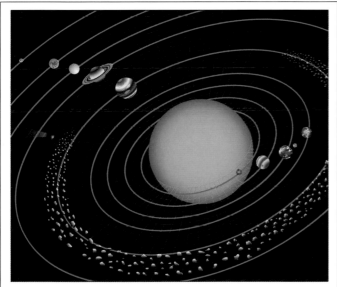

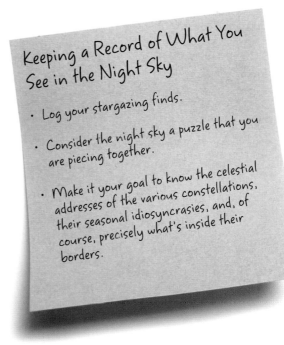

Keeping a Record of What You See in the Night Sky

- Log your stargazing finds.

- Consider the night sky a puzzle that you are piecing together.

- Make it your goal to know the celestial addresses of the various constellations, their seasonal idiosyncrasies, and, of course, precisely what's inside their borders.

- Newton's law of gravity enables us to calculate the masses of celestial objects, including the sun and the planets in our solar system, by assessing their orbiting habits.

- Understanding a little science goes a long way toward appreciating what you observe in space.

- Gravity is the force behind planets orbiting stars, like Earth orbiting the sun.

- Stars orbit galactic centers; galaxies orbit centers of mass in clusters of galaxies.

ASTROPHOTOGRAPHY

Photographing celestial objects is possible with reasonably priced equipment

Astrophotography used to be the province of a select few: professional astronomers and those with the financial wherewithal to purchase expensive equipment. But the times have changed, and now astrophotography is readily accessible to the everyday hobbyist. This is the case because many standard and affordable 35mm single reflex cameras

(SRCs), digital cameras, and economical astral cameras can handle basic astrophotography tasks. The days of every astrophotography undertaking employing a camera mounted at the end of a sizeable telescope are over. Amateurs can get in the game without Byzantine marriages of cameras and telescopes.

Astrophotography Camera

- Because of the long exposures required, film is better than digital technology for astrophotography purposes.

- Film can more readily handle minimal light conditions.

- A reasonably priced SLR (single lens reflex) camera is

a fine camera to begin your astrophotographic endeavors. These cameras not only facilitate long exposures but can often be attached to telescopes.

- You should always work with a camera that gives you the option to select a shutter speed.

Digital Astrophotography Camera

- The benefit of a digital camera in astrophotography is that poor quality pictures cost you nothing.

- Digital cameras also supply you with instantaneous results, which enable you to learn faster.

- Long exposures often cause

problems with digital cameras because of the heat generated in the picture-taking process.

- The moon is a recommended first target. But try to avoid it at its brightest. Photograph it several days before or after its full moon phase.

Astrophotography is at once simple and complex. For starters, with Earth always rotating, objects in the sky are in perpetual motion, even if their movements aren't apparent to the naked eye. More advanced astrophotography employs lengthy or multiple exposures to compensate for this celestial merry-go-round. And special filters offset obscuring pollutants and human-generated lighting, often the bane of stargazing. In addition, film emulsions that work with the minimal lighting of the night can augment your astrophotography.

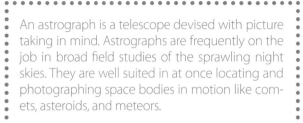

ZOOM

An astrograph is a telescope devised with picture taking in mind. Astrographs are frequently on the job in broad field studies of the sprawling night skies. They are well suited in at once locating and photographing space bodies in motion like comets, asteroids, and meteors.

Camera Accessories

- Some very basic camera accessories can make quality astrophotography possible.

- Look for specific adapter pieces that will work with your brand of camera. There are T-mounts and T-rings that can happily marry your camera and telescope.

- Pay heed to photographic trappings like remote releases and such, which open your camera's shutter automatically.

- Strive for minimal camera interferences and unnecessary vibrations in all your astrophotographic undertakings.

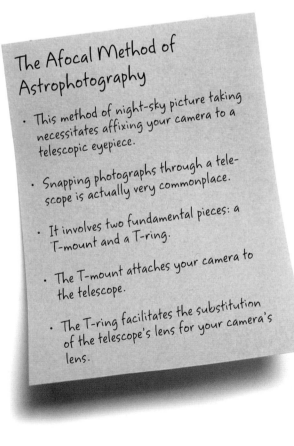

The Afocal Method of Astrophotography

- This method of night-sky picture taking necessitates affixing your camera to a telescopic eyepiece.

- Snapping photographs through a telescope is actually very commonplace.

- It involves two fundamental pieces: a T-mount and a T-ring.

- The T-mount attaches your camera to the telescope.

- The T-ring facilitates the substitution of the telescope's lens for your camera's lens.

23

NIGHT SKY FATHERS
Modern astronomy owes a debt of gratitude to its pioneers

Nicolaus Copernicus is sometimes called the "founder of modern astronomy." He was, in fact, the first scientist to publicly digress from the long-held view that planet Earth is a steady, unmoving mass at the nucleus of an enduring universe. Copernicus correctly surmised that Earth completely rotates around the sun in one year's time. He also believed that it fully rotates on its axis each day. Worth noting is that Copernicus's conclusions were entirely the byproduct of naked-eye observations. He had no amplifying instruments to assist him.

Galileo Galilei, who cast his lot with Copernicus's view that the sun—and not Earth—is the epicenter of our solar system, is similarly revered for his scientific contributions to astronomy and his pioneering grasp of the goings-on in the celestial sphere. Not only is Galileo the so-called father of contemporary physics, but also he is the man who put the

Nicolaus Copernicus

Galileo Galilei

- Nicolaus Copernicus's *De Revolutionibus Orbium Coelestium*, published in 1543, challenged the long-held view that Earth is an immobile orb and center of the universe.

- Copernicus was the first to commingle mathematics, physics, and cosmology

- with astronomy.

- He long resisted publication of his sun-centered heliocentric theory of the cosmos for fear of the consequences.

- Copernicus also studied economics, canon law, and medicine.

- Galileo Galilei put the telescope on the map.

- To survey the night skies, Galileo built a vastly advanced refractor telescope from an oral description of the newfangled instrument.

- Galileo's observations con-

- vinced him of the validity of Copernicus's heliocentric model.

- In 1633 the infamous Inquisition insisted that he publicly disavow Copernicus. He was convicted of heresy by the church and spent his last years under house arrest.

telescope on the map—or, in this instance, the star map. Galileo used a refractor telescope to explore the night skies and uncover never-before-known space bodies, including some of planet Jupiter's moons.

German scientist Johannes Kepler also deserves his due for supplying the fledgling science of astronomy with a solid point of reference. Kepler noted that the planets in our solar system move in ellipses around the sun. His "Laws of Planetary Motion" were at once groundbreaking and accurate.

Finally, Isaac Newton, a mathematician and physicist, merits recognition for his significant role in advancing the study of the celestial beyond. Frequently associated with the anecdote of an apple falling from a tree and simultaneous discovery of the law of gravity, Newton was ahead of his time. He understood that comparable gravitational influences are at play in space, governing the motion of planets and moons. He correctly surmised that gravity is the force holding them in their orbital places.

Johannes Kepler

- Johannes Kepler embraced the Copernican view of the cosmos.

- Kepler, however, recognized that the planets orbit the sun in ellipses, not circles. Kepler's three laws of planetary motion are still regarded as right on the mark, beginning with law 1:

"Planets move around the sun in ellipses, with the sun at one focus."

- Kepler's laws were not embraced with open arms. It wasn't until after his death that he was awarded his due.

Isaac Newton

- In 1687 Isaac Newton published *Principia*, which discoursed, among many things, on the mathematics of the known planets' orbiting patterns.

- Newton was the first person to calculate the relative masses of various celestial bodies—that is, exactly what mass it would take to command their respective gravitational patterns.

- Newton was also the first scientist to explain tidal ebbs and flows.

- Ironically, Isaac Newton was born the same year that Galileo died: 1642.

THE CELESTIAL SPHERE

This imaginary sphere plays an important role in understanding how the night sky works

The celestial sphere is a navigational tool that astronomers use to comprehend what's transpiring in the night sky. And you don't have to be a scientist to appreciate the celestial sphere's value in interpreting the seeming movements of space objects. For starters: The celestial sphere does not literally exist. It is a series of invented delineations, including a celestial equator

slicing across its midsection, the north celestial pole at its apex, and the south celestial pole identifying its nadir.

The celestial sphere is a globular illustration that enables you to visualize the impact of Earth's rotation on objects in the night sky that appear and eventually disappear. Most stars, for instance, move from east to west across the celestial sphere.

Imaginary Friend

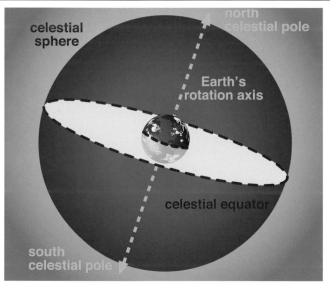

Right Ascension and Declination

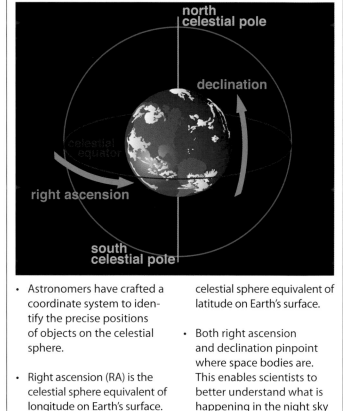

- The imaginary celestial sphere is rooted in three fundamentals: (1) extending Earth's equator outward forming the celestial equator; (2) extending one geographic pole upward forming the north celestial pole; (3) extending one geographic pole downward forming the south celestial pole.

- Earth is always at the center of the celestial sphere.

- At any given moment in time, you can see no more than one-half of the celestial sphere.

- Astronomers have crafted a coordinate system to identify the precise positions of objects on the celestial sphere.

- Right ascension (RA) is the celestial sphere equivalent of longitude on Earth's surface.

- Declination (DEC) is the celestial sphere equivalent of latitude on Earth's surface.

- Both right ascension and declination pinpoint where space bodies are. This enables scientists to better understand what is happening in the night sky and where objects are in relation to one another.

Their apparent motion takes them around the two celestial poles every night. This is not because they are soaring through outer space at warp speed. Quite the contrary: Earth's continuous rotation on its axis is the wind beneath their wings.

Paradoxically, Earth's rotation is west to east—and this counterclockwise direction is how it appears looking down from space at the North Pole. But here on Earth, we feel stationary and thus envision sky movements in reverse. And because our planet completely rotates on its axis each day, the fundamental night-sky dynamics remain in place every evening.

Polaris on the Celestial Sphere

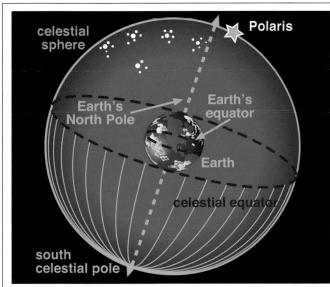

- In the imaginary celestial sphere, it is sometimes helpful to affix the North Star, Polaris, atop it.

- Polaris is less than 1 degree off the north celestial pole, which puts it pretty much at the apex of the celestial sphere.

- Earth's rotation axis always points to the north celestial pole and, by extension, Polaris.

- Polaris wasn't always the North Star. In 2000 B.C., Thuban in Draco had the distinction. In A.D. 14,000, Vega in Lyra will wear the crown.

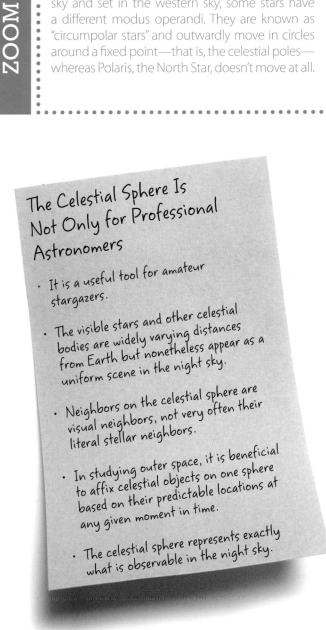

The Celestial Sphere Is Not Only for Professional Astronomers

- It is a useful tool for amateur stargazers.

- The visible stars and other celestial bodies are widely varying distances from Earth but nonetheless appear as a uniform scene in the night sky.

- Neighbors on the celestial sphere are visual neighbors, not very often their literal stellar neighbors.

- In studying outer space, it is beneficial to affix celestial objects on one sphere based on their predictable locations at any given moment in time.

- The celestial sphere represents exactly what is observable in the night sky.

APPRECIATING SIZE & SCALE

Our solar system is a fragment in a galaxy, one of billions, in a universe

The sun is one hundred times the size of our home, planet Earth. If a 747 jet airliner, traveling at its normal speed of 600 miles per hour, headed for the sun, it would take seventeen years to get there. Yet, this same sun, which sustains all life on this planet of ours, is, as stars go, rather average-sized.

To further add to this equation: There are an estimated two hundred billion to four hundred billion stars in our home galaxy alone. The Milky Way, where our solar system resides in the tiniest niche of its minor Orion arm, is just one of billions of galaxies in the sprawling universe at large.

To date, more than one hundred billion galaxies have been identified in what is often referred to as the "observable

Spectacular Sunset

- The sun is not any bigger when it's rising or setting on the horizon. This is another one of nature's optical illusions.

- The sun is one of hundreds of billions of stars in the Milky Way. Although it is not an unusual star, astronomers nonetheless surmise that it is among the top ten percentile in total mass.

- NASA's Hubble space telescope recently discovered a star that once shone with one million times the sun's brightness and weighed one hundred times its mass.

Earth versus the Moon

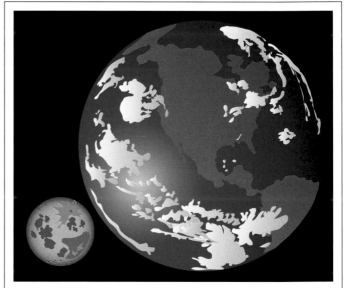

- Our moon is among the largest moons in our solar system. Still, it is only slightly more than one-quarter Earth's diameter at 2,160 miles (3,474 kilometers).

- The moon's surface area is less than one-tenth of our planet's surface area.

- *Apollo 11* took just a little over three days from blast-off to reach lunar orbit.

- A human mission to Mars would take approximately four months each way.

universe." In other words, this is what we absolutely know is in the celestial ether. It's what we don't yet know that's even more intriguing. Current observational estimates calculate that the universe is—at bare minimum—more than 150 billion light-years in width. Exactly how much more space and time are beyond our current capacities to detect? That's the big and unanswerable question that will likely always confound us.

And although Earth is the largest of the terrestrial planets closest to the sun, it is downright minuscule when compared with the planet Jupiter. It is estimated that approximately one thousand Earths could fit into Jupiter and that one thousand Jupiters could fit inside the sun. Yes, the same sun that is an average star when it comes to body mass. And consider this illuminating celestial measurement: The closest star, outside of the sun, to Earth is Proxima Centauri at 4.2 light-years away. In other words, it takes four years plus for that star's light to become visible here on Earth. Suffice it to say, there's a whole lot of space in outer space!

Dynamic Solar System

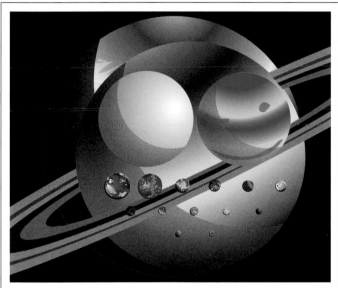

- Our sun has a mean radius of 432,000 miles (696,000 kilometers). No other planet or space body in our solar system comes close. Jupiter, the second-largest object at 43,440 miles (69,911 kilometers), is a puny gas ball relative to the sun.

- Our solar system consists of eight officially designated planets, three dwarf planets, and at least 130 natural satellites.

- There are likely many more objects to be discovered in the vast interplanetary medium of our solar system.

View from the Past

- The Hubble space telescope has supplied us with images of celestial objects and space neighborhoods as old as the presumed age of the universe.

- These snapshots furnish us with baby pictures of our young universe.

- The images reveal thickets of shapeless, simple matter that will eventually form entire galaxies of stars.

- If our universe is 13.7 to 14 billion years old, light from distances more than 14 billion light-years away has not yet reached us.

LIGHT-YEARS AWAY
When traditional measuring sticks fall short in outer space, light-years fill the vacuum

The human species prefers working with comfortable measurements: inches, feet, yards, and miles, for instance. So, when it comes to the sheer vastness of outer space, the comfort zone of earthly measuring sticks no longer applies. It's just not feasible to measure in miles or kilometers deep-space distances beyond our solar system. Scientists have thus devised a measurement known as "light-years" to fill this vacuum.

A light-year is the distance light travels in a year's time. And because light travels at 186,000 miles (300 kilometers) per second, a light-year is a considerable way from home. For example, it takes one hundred million years for the light from a space object 100 million light-years away to reach Earth.

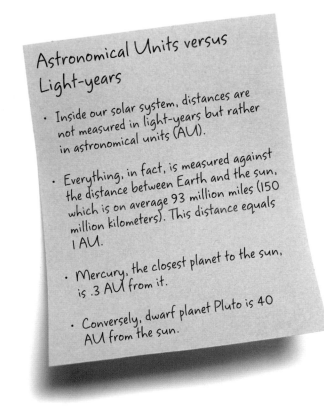

Astronomical Units versus Light-years

- Inside our solar system, distances are not measured in light-years but rather in astronomical units (AU).

- Everything, in fact, is measured against the distance between Earth and the sun, which is on average 93 million miles (150 million kilometers). This distance equals 1 AU.

- Mercury, the closest planet to the sun, is .3 AU from it.

- Conversely, dwarf planet Pluto is 40 AU from the sun.

Voyager 1

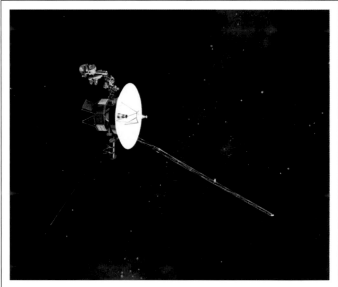

- The *Voyager 1* spacecraft was launched on September 5, 1977, and is still receiving commands from NASA scientists.

- Its flyby goal is reaching the heliopause, the suspected boundary of our solar system.

- Scientists anticipate future readings from *Voyager 1* regarding the ABCs of the interstellar medium.

- At present, *Voyager 1* is approximately 112 AU, or .0017 light-years away from the sun.

For comparative purposes here, the estimated time that it takes for sunlight reflected off the moon's surface to be observed on Earth is 1.2 to 1.3 seconds. Direct sunlight from the sun's surface takes 8.32 minutes to brighten the plane of Earth. But these measurements are from celestial bodies 238,000 and 93 million miles away, respectively. When we take into account that the nearest star to us, outside of the sun, is 24 billion miles away, you can appreciate why there's an absolute need to measure space objects in light-years rather than miles or kilometers.

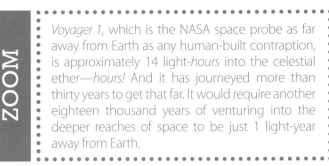

Voyager 1, which is the NASA space probe as far away from Earth as any human-built contraption, is approximately 14 light-*hours* into the celestial ether—*hours!* And it has journeyed more than thirty years to get that far. It would require another eighteen thousand years of venturing into the deeper reaches of space to be just 1 light-year away from Earth.

Triangulum Galaxy

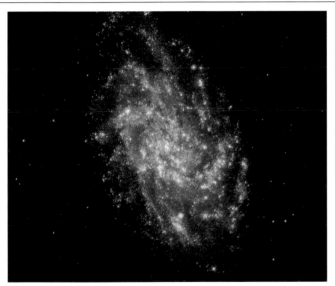

- The Triangulum galaxy, M33, is a spiral galaxy around 3 million light-years away in the northern constellation Triangulum.

- The Triangulum galaxy is small compared with the Milky Way and the Andromeda galaxies.

- Without light-years to measure its distance, it would be 13 quintillion miles (or 21 quintillion kilometers). That's 13,000,000,000,000,000,000 miles. Saying that our stellar neighbor is 2.3 million light-years away is a bit simpler.

Quasars

- Many scientists surmise that quasars represent the remotest areas in space: regions where new matter might very well be streaming into our universe. Quasars are, in fact, the most faraway objects yet discovered.

- Quasars, which are pointlike emanations of radio waves, are believed to shine with a trillion times the luminosity of our sun.

- Quasars are likely the active centers of galaxies.

- Most quasars span distances far larger than our entire solar system.

EARTH'S ATMOSPHERE
Earth's layers of atmosphere make life possible and distinguish it among planets

Earth would be unlivable without its several hundred miles of atmosphere. Indeed, our planet's unique atmosphere nicely multitasks as it soaks up energy from the sun and reprocesses water and other chemicals. Its magnetic force fields ensure a generally temperate climate that sustains life all across its widely varied surface.

Scientists have identified four distinctive layers of the atmosphere: the troposphere, stratosphere, mesosphere, and thermosphere. Beyond Earth's uppermost layer is an exosphere that bridges Earth's atmosphere with the more forbidding and uninhibited gases of outer space.

The troposphere, which begins at Earth's surface and ascends

Home Is Where the Troposphere Is

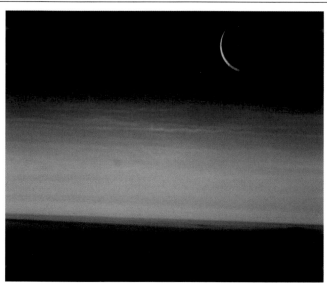

- The clouds in the sky are in the lowest layer of Earth's atmosphere: the troposphere. Because this layer begins at the surface, we, too, are in the troposphere.

- Air is warmest at the bottom layer of the troposphere—at ground level—where we are.

- Approximately 75 to 80 percent of our atmosphere's mass is found in the troposphere.

- The tropopause is a region between the troposphere and the next layer of Earth's atmosphere: the stratosphere.

Stratospheric Clouds

- Although few clouds are found in the layer of atmosphere known as the "stratosphere," polar stratospheric clouds (PSCs) are the exception.

- PSCs are visible near the two poles during wintertime and are sometimes called "nacreous clouds."

- Commercial jets fly in the lower portion of the stratosphere to avoid the turbulence and weather of the lower atmosphere.

- Air is thin in the stratosphere, approximately one thousand times thinner than the air we breathe.

some 5 to 9 miles (8 to 14.5 kilometers), is indisputably the densest of all the layers of the atmosphere. It is also where our weather events play out. The stratosphere assumes the baton immediately above the troposphere. It is a less dense and drier layer which climbs 31 miles (50 kilometers). Here, too, is where Earth's ozone layer, which grabs hold of and disperses solar ultraviolet radiation, resides. Without the ozone layer providing this welcome service, we would be toast—literally.

The mesosphere is the third layer of atmosphere and spans 53 miles (85 kilometers). Here the temperatures fall dramatically, and chemicals abound in highly agitated states. The thermosphere, which includes the overlapping ionosphere, is referred to as Earth's "upper atmosphere" and bucks the mesosphere's falling temperature trend. In the thermosphere, temperatures increase with the rising altitude, which brings one closer to the sun's formidable heat force and solar wind. The thermosphere soars 372 miles (600 kilometers).

Mesospheric Clouds

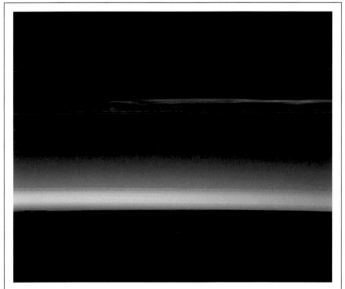

- Noctilucent clouds can be seen in the mesospheric layer of Earth's atmosphere close to both the North Pole and South Pole.

- More than any other cloud types in Earth's multiple layers of atmosphere, these are categorically the highest forming clouds.

- The upper portion of the mesosphere is the coldest layer of atmosphere—colder than even the thermosphere above it.

- The majority of meteors go up in smoke, so to speak, in the mesosphere.

Space Weather

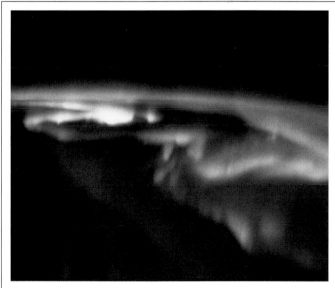

- The aurorae—both the Northern Lights and Southern Lights—perform in the thermosphere.

- Outer space particles colliding with atoms and molecules inspire aurorae.

- Both the space shuttle and the International Space Station orbit our planet within the thermosphere.

- The ionosphere, which embodies regions of the upper atmosphere with magnetized particles, shares ether with the thermosphere. It is not considered a separate layer of atmosphere.

EARTH'S ROTATION

Earth's rotation on its axis and orbit around the sun define our days and nights

Although we don't literally feel it, our home planet is in continual motion, simultaneously rotating on its axis and orbiting the sun. And while we don't sense Earth spinning in any way or suspect we are on an annual jaunt around the sun, we indisputably see the evidence before us.

It actually takes Earth 23 hours, 56 minutes, and 4.09 seconds to make a complete revolution on its axis. At the equator, Earth's rotating speed has been calculated at 1,070 miles (1,723 kilometers) per hour. This daily show, starring Earth, is what furnishes us with day, then night, and then day again. The unfailing rising of the sun in the east and setting in the west are courtesy of the Earth's rotation.

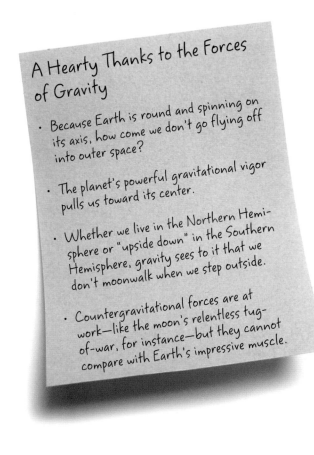

A Hearty Thanks to the Forces of Gravity

- Because Earth is round and spinning on its axis, how come we don't go flying off into outer space?

- The planet's powerful gravitational vigor pulls us toward its center.

- Whether we live in the Northern Hemisphere or "upside down" in the Southern Hemisphere, gravity sees to it that we don't moonwalk when we step outside.

- Countergravitational forces are at work—like the moon's relentless tug-of-war, for instance—but they cannot compare with Earth's impressive muscle.

Off Kilter

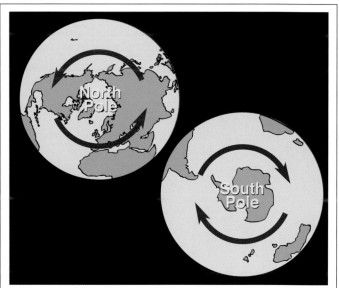

- Earth's rotation on its axis is not perpendicular but rather tilted some 23.5 degrees vis-à-vis the plane of the ecliptic.

- Earth's inclination ensures that at various points in its orbit around the sun, regions of the planet are skewed more toward it.

- And, conversely, regions are also tipped away.

- Earth's tilting on its axis is why we have seasons.

- From the vantage point of outer space, Earth rotates on its axis in a counterclockwise direction.

It is also Earth's unceasing movement that welcomes the star-filled night skies each evening, with the preponderance of the stars seemingly rising—just like the sun—in the east and setting in the west. This recurring behavior of the night skies is known as "diurnal motion." The celestial bodies that come into view and inexorably glide across the night sky are, in reality, in fixed locations in space. Their changing position as the night wears on is, again, the consequence of the third rock from the sun's spinning on its axis.

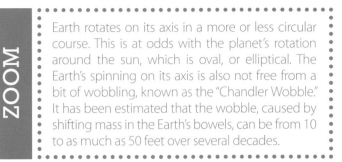

ZOOM

Earth rotates on its axis in a more or less circular course. This is at odds with the planet's rotation around the sun, which is oval, or elliptical. The Earth's spinning on its axis is also not free from a bit of wobbling, known as the "Chandler Wobble." It has been estimated that the wobble, caused by shifting mass in the Earth's bowels, can be from 10 to as much as 50 feet over several decades.

Cold Spring Sunrise

- Beginning in springtime, the sun climbs incrementally higher in the sky.

- The warmer seasons of late spring and summer see the sun above the horizon for longer periods of time than do autumn and winter.

- As summer approaches, the sun's rays wallop the surface of Earth more directly than during the other seasons of the year.

- The tilt in Earth's rotation on its axis governs the solar radiation—warming and lack thereof—that we enjoy on the surface.

Winter Sunset

- The sun is above the horizon for shorter spans of time during the winter.

- The sun's rays affect the surface of Earth more indirectly during wintertime than summertime.

- Ironically, Earth is literally nearer to the sun during

Northern Hemisphere and Southern Hemisphere winters than during their respective summers. It is the axis-tilt that is definitive. The 23.5-degree tilt of Earth's axis, which supplies us with seasons, is known as the "obliquity" of Earth's axis.

OUR HOME GALAXY

We live on a planet inside a solar system inside a galaxy: the Milky Way

Quite often the most spectacular visual in the night sky is our parent galaxy: the Milky Way. The Milky Way is classified as a spiral galaxy because of its presumed shape and features. It hosts our solar system in one of its minor spiral arms—or spurs—known as the "Orion arm."

The Milky Way appears as a creamy but nonetheless vivid swath of light arcing across a healthy portion of the night sky. The entire galaxy accommodates an estimated two hundred billion to four hundred billion stars, nebulae, and a range of interstellar gases and dust clouds. The merging light emissions from these many faraway space objects are what supply the Milky Way with its distinctive sheen.

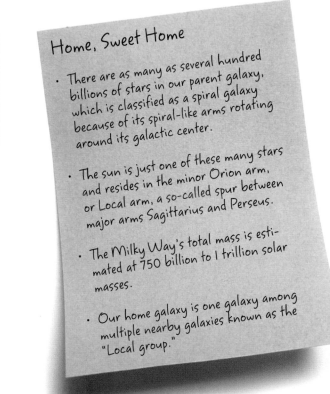

Home, Sweet Home

- There are as many as several hundred billions of stars in our parent galaxy, which is classified as a spiral galaxy because of its spiral-like arms rotating around its galactic center.

- The sun is just one of these many stars and resides in the minor Orion arm, or Local arm, a so-called spur between major arms Sagittarius and Perseus.

- The Milky Way's total mass is estimated at 750 billion to 1 trillion solar masses.

- Our home galaxy is one galaxy among multiple nearby galaxies known as the "Local group."

Milky Way: Up, Up, and Away

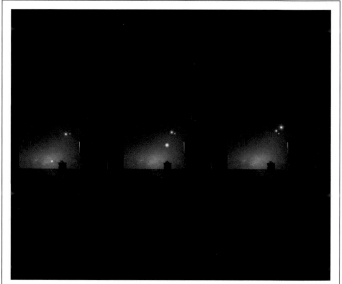

- The Milky Way got its name courtesy of the dim but nonetheless distinctive band of light that arcs across the celestial sphere.

- The light is the consequence of billions of bright stars residing along its galactic plane.

- The center of the Milky Way is located in the vicinity of the constellation Sagittarius.

- The galaxy's ultrabright nucleus is believed to accommodate very old stars, which enlarge and shine brightest at the ends of their celestial lifetimes.

Ironically, an ongoing mystery surrounds the galaxy that includes planet Earth and the sun as tenants. Because it is impossible to completely study, measure, and photograph something that we are within, the Milky Way remains a somewhat unknown astronomical quantity. We can see only portions of it from our vantage point—an insider's view, so to speak. Therefore, we can never get a completely accurate picture—top to bottom, side to side—of the galaxy that houses our solar system within its vast boundaries.

Astronomers, nevertheless, have a handle on the critical parts of the Milky Way. Foremost, its disk shape accommodates innumerable interstellar bodies and space matter. This disk also includes the galaxy's assumed spiral arms: Perseus, Sagittarius, Norma, and Scutum-Centaurus. The Milky Way also sports spherical components known as the "halo" and "central bulge." The bulge, or nucleus, is the all-important galactic center of mass, which holds together, via gravity, the conglomeration of celestial objects in the galaxy. A potent galactic center is, in fact, what makes a galaxy a galaxy. This gravitational epicenter is also where a mysterious black hole is believed to exist.

Billions of Stars' Sheen

- The "fuzzy swath of light" that characterizes the Milky Way is the byproduct of billions of stars' incandescence.

- A vast amalgam of light sources from disparate distances fashions a homogeneous "milky" appearance on the celestial sphere.

- The galaxy's considerable dust further diffuses its radiant swath of light. The wide disparity in total star estimates of the galaxy-comes from the fact that most of its stars are low mass and cannot be seen.

Neighboring Large Magellanic Cloud

- The Large Magellanic Cloud was once considered a Milky Way satellite galaxy.

- Both the Large Magellanic Cloud and its companion, the Small Magellanic Cloud, interact with the Milky Way.

- These two irregular galaxies, with amorphous contours, are visible only in the Southern Hemisphere.

- Scientists believe that these two nearby galaxies constrict the size of the Milky Way's galactic halo, which reaches outward both above and below its plane.

SPIRAL ARMS

Our home galaxy's spiral arms rotate around its robust galactic center

Part of both the Milky Way's inscrutability and its idiosyncratic charm is its so-called spiral arms. It's our home galaxy's apparent space arms jutting approximately 10,000 light-years away from its center that supply it with its suspected shape and, hence, designation as a spiral galaxy. That is, these arms housing stars, including our sun, and interstellar matter revolve—or spin, if

you will—around the Milky Way's formidable galactic center.

Not surprisingly, the precise nature of the Milky Way's spiral arms is still a matter of conjecture. In fact, the scientific consensus of these arms has changed through the years. Originally, it was believed that our galaxy has more spiraling arms than are currently surmised, including an Orion arm, home of our solar

Major and Minor Spiral Arms

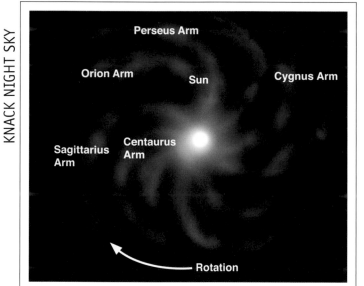

Perseus Arm

Orion Arm

Sun

Cygnus Arm

Sagittarius Arm

Centaurus Arm

Rotation

Milky Way Doppelganger

- The number, precise shapes, and movements of the Milky Way's spiral arms have long been the subject of scientific evaluation and reevaluation.

- The Cygnus arm is alternatively known as the Norma arm; the Centaurus arm, the Scutum-Centuarus arm.

- A consensus seemed to have developed that the Milky Way has four major spiral arms: Perseus, Norma, Scutum-Centaurus, and Sagittarius. Recent findings intimate that the galaxy may have only two major arms: Perseus and Scutum-Centaurus.

- Because we can never get a complete portrait of our parent galaxy, scientists rely on visual images from distant galaxies similar to our own.

- NGC 1097 is a barred spiral galaxy with a configuration believed to be comparable with the Milky Way's design.

- NGC 1097 is approximately 45 million light-years away.

- Located in the constellation Fornax, NGC 1097 exhibits a highly energetic galactic center likely under the sway of a massive black hole.

system. It is now widely felt that the Orion arm is merely a minor arm—a spur between major arms Sagittarius and Perseus.

However, recent discoveries from NASA's Spitzer space telescope, which is capable of detecting infrared light, have uncovered objects and matter in our galaxy never before observed and deciphered. All space matter releasing heat can be detected in infrared. These fresh findings have made the Milky Way's wholly gaseous spiral arms—previously detectable via only radio wavelengths —more transparent.

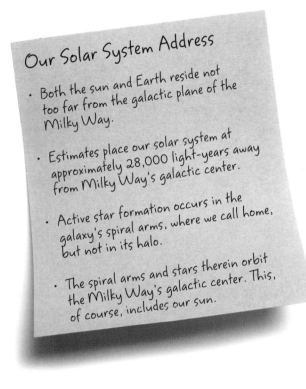

Our Solar System Address

- Both the sun and Earth reside not too far from the galactic plane of the Milky Way.

- Estimates place our solar system at approximately 28,000 light-years away from Milky Way's galactic center.

- Active star formation occurs in the galaxy's spiral arms, where we call home, but not in its halo.

- The spiral arms and stars therein orbit the Milky Way's galactic center. This, of course, includes our sun.

We Are Here

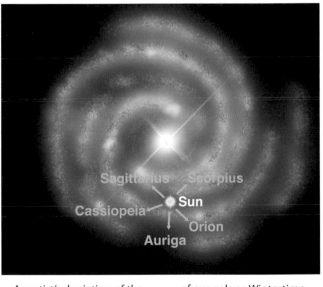

- An artist's depiction of the Milky Way identifies our sun and solar system's position. The Milky Way's spiral arms rotate around its pronounced bulge and center bar of concentrated stars.

- During the summer months, we face Sagittarius and the bright nucleus of our galaxy. Wintertime places us at a wholly different angle, toward Auriga, and away from the radiant core.

- Our little niche in the Milky Way is considered the most hospitable region in the entire galaxy.

THE MILKY WAY

39

DISK OF STARS & DUST

The Milky Way's disk is its fundamental characteristic and what we observe in the night sky

When we can observe the sprawling Milky Way contours in the night sky, we are actually seeing the galaxy's stellar disk. Indeed, the Milky Way's elongated disk shape is its defining feature. And, here again, it's impossible to fully comprehend the makeup of this disk. Because we are residents of the disk's interior, the inability to survey it in its entirety is an ongoing scientific challenge. Nevertheless, what astronomers can absolutely say is that the disk is home to multiple billions of stars, red-hot nebulae, and stone-cold dust filling in the vast nooks and crannies between and among celestial bodies. This considerable cosmic space, known as the "interstellar medium," erects a further barrier to the study of the

Defined Disk Shape

- An extensive disk, spanning 100,000 light-years from end to end, encircles the Milky Way's bar-shaped midsection.

- The galactic disk's intense brightness is greatly diffused by the immense interstellar medium.

- The Milky Way's disk is chock full of gases and dark clouds of dust.

- The galactic disk is encircled by a spherical-shaped halo, as it's called, which is a sanctuary for older stars and globular clusters.

Summer Milky Way

- The summer night-sky Milky Way climbs high in the north sky and spans the constellations Cygnus and Aquila.

- To affix your observing eyes near the galactic center, gaze across the southern horizon into the constellations Sagittarius and Scorpius.

- The Great Rift—a dark band in the Milky Way shaped by dense interstellar clouds—is also evident in the summertime.

- The Great Rift's murky band runs between the constellations Cygnus and Scutum.

Milky Way. The aforementioned space dust cloaks immense regions of the galaxy, making it difficult to gauge what's happening behind the opaqueness.

In fact, the majority of the brightest stars in our galaxy, including the sun, can be found within the disk. The disk also contains the Milky Way's various spiral arms. The broadest and most active quarter of the disk is located at its mid-section. This is the galaxy's nucleus and the center of gravity. Here a gravitational gravitas binds together the sum total of the galaxy.

Winter Milky Way

- The winter night sky offers a much darker portrait of the Milky Way.

- The Milky Way is higher in the sky and positioned away from its bright galactic center.

- Winter night viewings of the Milky Way are nonetheless worthwhile endeavors. The crisp, clear winter air, commingling with the darker galaxy higher overhead, furnishes observers with unmistakable visual contrast.

- Winter is the season with lower humidity and better air quality all around.

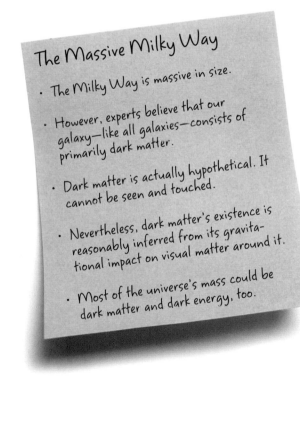

The Massive Milky Way

- The Milky Way is massive in size.

- However, experts believe that our galaxy—like all galaxies—consists of primarily dark matter.

- Dark matter is actually hypothetical. It cannot be seen and touched.

- Nevertheless, dark matter's existence is reasonably inferred from its gravitational impact on visual matter around it.

- Most of the universe's mass could be dark matter and dark energy, too.

BULGE
The Milky Way's galactic center visibly and brilliantly protrudes from its disk shape

The markedly bright area of the Milky Way's prominent bend across the rambling night sky is its galactic center, or bulge, as it's frequently called. This so-called bulge is the spherical heart and soul of our galaxy. The Milky Way's nucleus, which noticeably protrudes out from its prevailing disk, accommodates a thick base of stars orbiting around its gravitational center.

Although we cannot observe the sum of the Milky Way, a widely held view is that it is shaped like a disk: flat and rounded like a pancake but with a pronounced elliptical protuberance at its midpoint. Courtesy of its well-defined brightness, it is indisputably a region of the galaxy teeming with interstellar commotion. And this cosmic snapshot is

Active Space Neighborhood

Central Bulge in Infrared

- The Milky's Way bulge is slightly extended in the direction of the sun.

- Whereas most regions of our galaxy are vast stretches of emptiness, the central bulge is teeming with activity.

- The nearest star to our sun is 4.2 billion light-years away, yet there are approximately ten million stars in a 1 light-year area close to the central bulge.

- The bulge houses mostly old stars believed to be almost as old as the galaxy itself.

- Infrared images of the central bulge clearly reveal this region's hyperactivity and commensurate brightness.

- Stars in the central bulge are unlike the star population residing in the outermost areas of the disk, including the spiral arms.

- There are many aged red giant stars in the galactic bulge.

- This fact underscores the significant genetic differences in the Milky Way's stars based on their location.

evident not only to scientists with high-powered telescopes but also to the countless naked eyes combing the celestial sphere at any given moment.

Although astronomers readily acknowledge the Milky Way's confounding character, they nonetheless accept the fact that its conspicuous bulge includes many mature stars that have birth dates that coincide with the galaxy's presumed age. These intriguing findings are invaluable to scientists seeking greater insight into the origins of our parent galaxy.

Indeed, the Milky Way's bulge is home to stars ten billion years or older. And although the precise age of our galaxy is hard to pin down, it is widely felt to be 13.7 billion years old or close to it. Of course, black holes are what many scientists trust bind together all galaxies, including our Milky Way. That is, areas in outer space with such physically powerful gravitational muscle that nothing—absolutely nothing—can escape their considerable and mystifying tugs. If the Milky Way didn't contain a super-massive black hole at its core, many astronomers believe that the luminescence from its present emanations would be more than 250 times brighter—and that's saying something.

Supernovae: Black Hole Makers

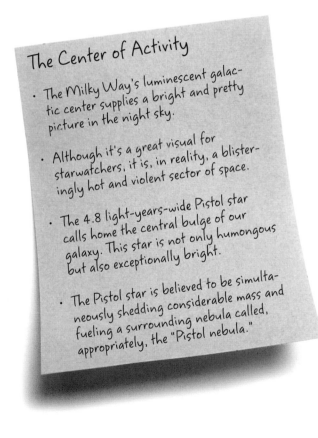

The Center of Activity

- The Milky Way's luminescent galactic center supplies a bright and pretty picture in the night sky.

- Although it's a great visual for starwatchers, it is, in reality, a blisteringly hot and violent sector of space.

- The 4.8 light-years-wide Pistol star calls home the central bulge of our galaxy. This star is not only humongous but also exceptionally bright.

- The Pistol star is believed to be simultaneously shedding considerable mass and fueling a surrounding nebula called, appropriately, the "Pistol nebula."

- Likely lurking in the recesses of the Milky Way's bright and formidable galactic center is a massive black hole.

- Not all spiral galaxies have central bulges of stars like the Milky Way.

- A recent finding of the galaxy's rotation speed—568 million miles (914 kilometers) per hour—reveals that it is moving much faster than earlier surmised.

- This discovery means the Milky Way is much more massive than previously thought. It also means that collisions with nearby galaxies are in the offing.

HALO

The Milky Way wears a gamma-ray halo consisting of mysterious dark matter

On either side of the Milky Way's elongated galactic plane is a puzzling gamma-ray halo of sorts. This considerably larger and more sphere-shaped offshoot of our home galaxy, which stretches approximately five thousand to 10,000 light-years above and below its radiant disk, is believed to hold important clues vis-à-vis the Milky Way's formation and subsequent growth. And considering that our galaxy's disk-like contour spans roughly 100,000 light-years in total distance, the halo is something scientists are very interested in.

Indeed, like all other features of the Milky Way, its halo is shrouded in some authentic mystery. In fact, the halo effect—found in other galaxies as well—that rises above and

Halo of Stars

- The brightest objects in the Milky Way's halo are globular clusters.

- The perceptible stars in the halo represent only a minuscule proportion of the mass in this region of the Milky Way.

- Dark matter is believed to govern the halo.

- Scientists recently described its surrounding halo as probably resembling a "squashed beach ball." That is, if this mysterious dark matter would reveal itself.

Ultrabright Halo

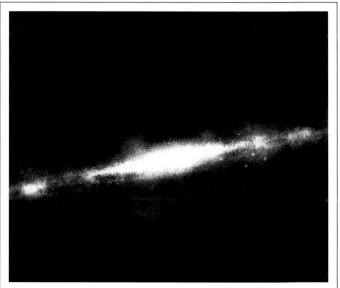

- The brightest regions of the halo may contain remnants of past galaxies that were gobbled up by the more massive Milky Way.

- Many astronomers suspect that lesser galaxies have, in fact, merged with our parent galaxy over time.

- If this is indeed what we are seeing, the most luminescent sectors of the halo could be former nuclei of once-independent galaxies.

- This process—known as "accretion"—not only happened once upon a time but also is happening in the present.

below the Milky Way like a ghostly colossus is likely the product of a protracted series of exploding stars.

Although still under scientific scrutiny, the consensus is that the Milky Way exhibits a super-hot gaseous aura. It is widely believed that this inscrutable nimbus constitutes the lion's share of the galaxy's overall mass, too. But here's where it gets even more interesting: A halo within the Milky Way's visible halo is presumed to consist of dark matter, which is undetectable via standard radiation emissions and analysis.

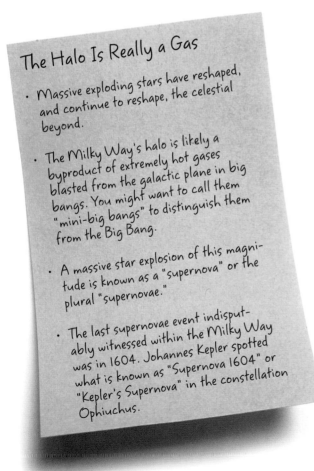

The Halo Is Really a Gas

- Massive exploding stars have reshaped, and continue to reshape, the celestial beyond.

- The Milky Way's halo is likely a byproduct of extremely hot gases blasted from the galactic plane in big bangs. You might want to call them "mini-big bangs" to distinguish them from the Big Bang.

- A massive star explosion of this magnitude is known as a "supernova" or the plural "supernovae."

- The last supernovae event indisputably witnessed within the Milky Way was in 1604. Johannes Kepler spotted what is known as "Supernova 1604" or "Kepler's Supernova" in the constellation Ophiuchus.

Supernova Explosion

- Supernovae—that is, exploding stars—are incredibly dramatic, colorful celestial events. Supernovae shake and bake the interstellar medium and cast heavy metals throughout space with unrestrained abandon.

- They are believed to be the underpinning of our galaxy's gaseous halo.

- What you see in the supernovae are multiple thousands of pointlike X-ray streams.

- The red colors indicate low X-ray emissions; the green, medium emissions; and the blue, high emissions.

ORION ARM
Our solar system resides in a minor arm, or spur, in the Milky Way

It is a common question, one that intrigues amateur astronomers and fledgling stargazers peering into outer space: *Where do we reside in our parent galaxy?* The short answer is within the Milky Way's Orion arm, a minor spiral arm in the overall galaxy. In fact, the region where our solar system dwells is sometimes called the "Local arm," "Local spur," or "Orion spur." More specifically, the sun and Earth inhabit an area of the galaxy between two of its major spiral arms: Perseus and Sagittarius.

Not too long ago, the Orion arm was considered a bona fide spiral arm in the galaxy at large. The fact that it has been demoted is indicative of increasingly in-depth scientific findings furnishing us a clearer picture of our galaxy's structure. Although the Orion arm is now considered a minor piece of the puzzle, it is nonetheless the site of our solar system, which houses both the planet we call home and the sun that nurtures and sustains most life forms on Earth.

Solar System in the Orion Arm

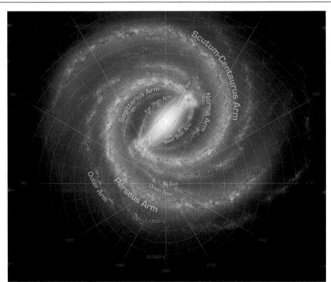

- The Orion arm—minor though it may be—is still 3,500 light-years wide and 10,000 light-years long.

- The sun, Earth, and solar system at large are situated near Orion arm's inner rim.

- More precisely, we call home an area very near the midpoint of the Orion arm.

- The Orion arm got its moniker because of its proximity to the constellation Orion.

Between Sagittarius and Perseus

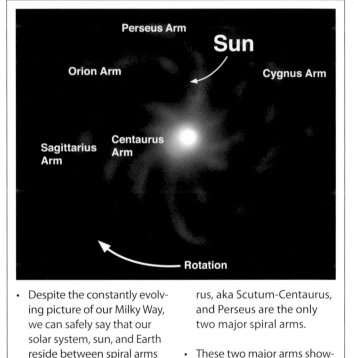

- Despite the constantly evolving picture of our Milky Way, we can safely say that our solar system, sun, and Earth reside between spiral arms Sagittarius and Perseus.

- The latest scientific scuttlebutt suggest that Centaurus, aka Scutum-Centaurus, and Perseus are the only two major spiral arms.

- These two major arms showcase the maximum densities of young, intense stars as well as older red giant stars.

The Orion arm, or spur, received its moniker because of its propinquity to the constellation Orion the Hunter. Our solar system in Orion is situated approximately 26,000 light-years away from the galaxy's galactic center. It is in what is called the "Local bubble," which is approximately halfway along the span of the arm.

And although the Orion arm has lost its vaunted status as a significant spiral arm in the Milky Way, it nonetheless harbors a plethora of popular space bodies first identified by French astronomer Charles Messier. In the eighteenth century Messier cataloged a wealth of deep-sky objects, which are still referenced today. The Orion arm of the Milky Way includes—among many others—the Butterfly cluster, Dumbbell nebula, Beehive cluster, Pleiades, and Ring nebula. These are dubbed "Messier's objects."

Just Passing Through

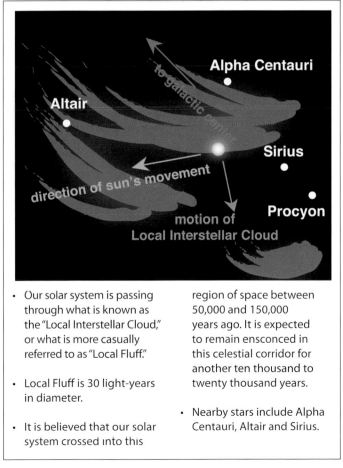

- Our solar system is passing through what is known as the "Local Interstellar Cloud," or what is more casually referred to as "Local Fluff."

- Local Fluff is 30 light-years in diameter.

- It is believed that our solar system crossed into this region of space between 50,000 and 150,000 years ago. It is expected to remain ensconced in this celestial corridor for another ten thousand to twenty thousand years.

- Nearby stars include Alpha Centauri, Altair and Sirius.

Local Bubble

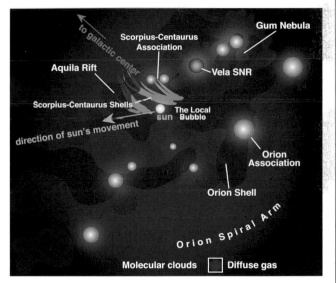

- The so-called Local Bubble, shaded dark blue, is an expansive area of space that our solar system has been journeying through for the last five to ten million years.

- The Local Interstellar Cloud calls home the Local Bubble.

- The Local Bubble is approximately 300 light-years in diameter. Star-forming regions are colored purple.

- It is suspected that this region of the interstellar medium in the Orion arm is the consequence of supernovae explosions.

THE NORTHERN HEMISPHERE

The Northern Hemisphere encompasses everything north of the equator, including the majority of the population

The term *hemisphere* simply means a "half sphere." More specifically, the Northern Hemisphere denotes the top half of our planet: the entire area of Earth that lies north of the equator. And for astronomical purposes, the Northern Hemisphere also embodies the so-called celestial sphere and all that lies above the celestial equator.

In the overall study of astronomy, the Northern Hemisphere indisputably gets more play than its southern counterpart. And this isn't because the night skies of the Northern Hemisphere are so much more impressive—or any more impressive, for that matter—than Southern Hemisphere night skies. It is sheer demographics. That is, a whole lot more people live in the north

Northern Hemisphere

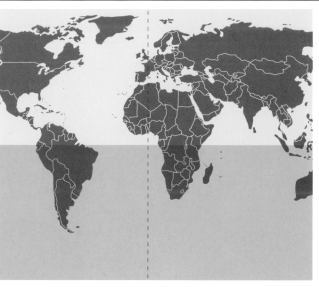

- The Northern Hemisphere encompasses the upper half of planet Earth—everything that is north of the equator.

- The map of the Northern Hemisphere clearly reveals how this portion of the planet contains the vast majority of landmass and,

- by extension, population.

- Even the northernmost reaches of South America lie in the Northern Hemisphere.

- Approximately two-thirds of the African continent is located in the Northern Hemisphere.

Earth's Upper Half

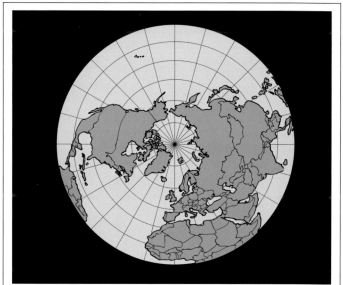

- All of North America and Europe are in the Northern Hemisphere. With the exception of parts of Indonesia, just about the entire continent of Asia is as well.

- The equator runs through some countries, including Ecuador, Colombia, Brazil, Gabon, Congo, Kenya,

- Somalia, and Maldives. These countries have areas lying in both hemispheres.

- The Arctic spans the region north of the Arctic Circle. Very cold winters and cool summers are the norm.

than do in the south. And, not surprisingly, with more accessibility to what's in their own backyards, so to speak, there is a corresponding interest in what can be observed with their own two eyes. Think about it: The Northern Hemisphere includes the entire continents of North America and Europe. Most of Asia and Africa are located in the Northern Hemisphere, as is a northern sliver of South America. To paint the most unambiguous picture possible: The Northern Hemisphere is home to approximately 90 percent of the world's population as well as the majority of its landmass.

Northern Hemisphere denizens and starwatchers behold night skies that are generally less bright with fewer visible stars than their peers to the south of the equator. This is courtesy of the Northern Hemisphere sky's position with respect to the Milky Way's bright galactic center. But this doesn't mean the night skies are any less striking or of any less interest to professional and amateur astronomers. In fact, the relatively darker night skies in the north enable stargazers to get a more intimate view of deep-sky objects in the far recesses of space.

Dark Night Sky

- Courtesy of the North Pole's surface location vis-à-vis the Milky Way's night sky positioning, our galaxy's galactic center faces away from the Northern Hemisphere.

- This celestial snapshot leaves the Northern Hemisphere more in the dark than the Southern Hemisphere.

- That is, the Northern Hemisphere night skies are conspicuously darker than their counterparts below the equator.

- There are fewer visible stars in the Northern Hemisphere night sky than in the Southern Hemisphere night sky.

Northern Skies Star Map

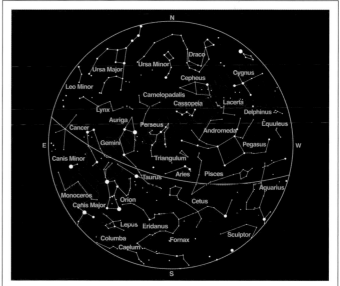

- The Northern Hemisphere has multiple constellations of stars that prominently reveal themselves during different seasons of the year.

- Canis Major, Orion, Perseus, and Taurus are all considered northern winter constellations. This means that they are high in the sky during wintertime and visible to stargazers throughout the Northern Hemisphere.

- The Northern Hemisphere's summer constellations include Cygnus, Hercules, Lyra, Sagittarius, and Scorpius.

DISTINCTIVE FEATURES
The Northern Hemisphere night sky is darker than its Southern Hemisphere counterpart

The Northern Hemisphere night skies are markedly different from the Southern Hemisphere night skies. But this doesn't mean that they are two wholly unique shows. Most space bodies visible in the Northern Hemisphere are also visible in the Southern Hemisphere, although at different times of the year and in different slices of the sky. There are, of course,

some stargazing attractions that are seen only in the respective hemispheres.

Courtesy of the Earth's daily rotation on its axis and annual orbit around the sun, your physical location therefore matters with respect to what you are seeing up above at any given moment. Think about it: A sundial moves clockwise in

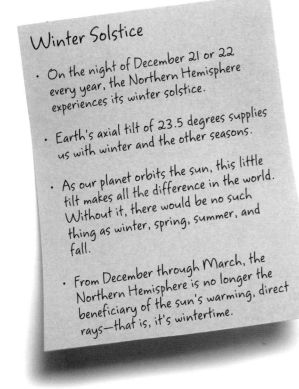

Winter Solstice

- On the night of December 21 or 22 every year, the Northern Hemisphere experiences its winter solstice.

- Earth's axial tilt of 23.5 degrees supplies us with winter and the other seasons.

- As our planet orbits the sun, this little tilt makes all the difference in the world. Without it, there would be no such thing as winter, spring, summer, and fall.

- From December through March, the Northern Hemisphere is no longer the beneficiary of the sun's warming, direct rays—that is, it's wintertime.

Cassiopeia the Queen

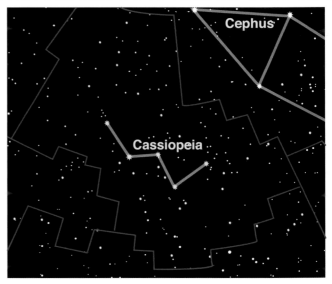

- Cassiopeia the Queen is an eminent northern sky constellation. During optimal viewing times, the constellation is readily spotted with its W- shaped five bright stars.

- Two Messier's objects are worth checking out in Cassiopeia: M52 and M103.

- These globular clusters can be observed with binoculars.

- Cassiopeia is yet another constellation named for a figure out of Greek mythology.

- Her daughter was Andromeda.

the Northern Hemisphere, counterclockwise in the Southern Hemisphere. And as the sun makes its daily trek across the sky—rising in the east and setting in the west in both hemispheres—its maximum position is to the south in the Northern Hemisphere and to the north in the Southern Hemisphere.

Paradoxically, the vastly more explored Northern Hemisphere night sky is not as well lit when contrasted to its southern cousin. In fact, it sports only five circumpolar constellations—that is, constellations that are observable all year long: Ursa Major, Ursa Minor, Cephus, Cassiopeia, and Draco.

And within these constellations is not one star deemed of first-order magnitude or higher. That is, there are no really bright stars in the mix. Conversely, the Southern Hemisphere accommodates eleven circumpolar constellations, with six stars of first-order magnitude.

That said, the Northern Hemisphere has a pole star: Polaris, aka the "North Star." This fairly bright star rests precisely atop the celestial North Pole. At no time can residents of the Southern Hemisphere ever glimpse the legendary North Star.

Cygnus the Swan

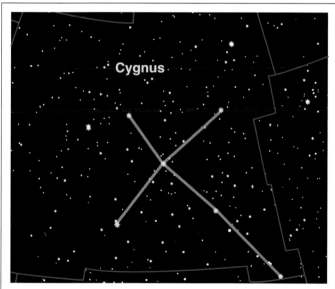

- Cygnus the Swan is a northern constellation renowned for accommodating the Northern Cross.

- The Northern Cross is an asterism of stars—not to be confused with the more famous Southern Cross—with Albireo, a double star, atop and Deneb its anchor.

- Deneb is also a key star in another familiar asterism: the Summer Triangle.

- Interestingly, the Summer Triangle consists of three stars from three constellations: Altair from Aquila, Deneb from Cygnus, and Vega from Lyra.

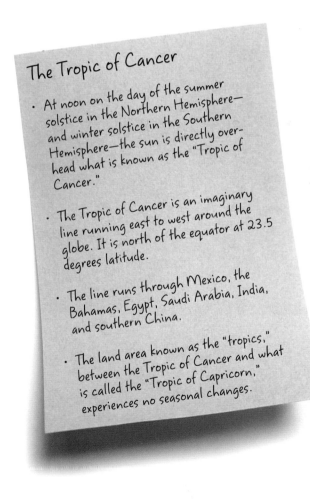

The Tropic of Cancer

- At noon on the day of the summer solstice in the Northern Hemisphere—and winter solstice in the Southern Hemisphere—the sun is directly overhead what is known as the "Tropic of Cancer."

- The Tropic of Cancer is an imaginary line running east to west around the globe. It is north of the equator at 23.5 degrees latitude.

- The line runs through Mexico, the Bahamas, Egypt, Saudi Arabia, India, and southern China.

- The land area known as the "tropics," between the Tropic of Cancer and what is called the "Tropic of Capricorn," experiences no seasonal changes.

WINTER NIGHT SKIES

Winter night skies in the Northern Hemisphere are crisp, clear, and brimming with sightings

The winter night skies in the Northern Hemisphere rival the presentation of the summer night skies. It's just that not as many people take advantage of both the longer nights and relatively clearer skies of wintertime to scope out the celestial beyond. Yes, it's cold out there at this time of year. But none-theless there's a lot to see in the night sky during the months of December, January, February, and into March.

The Northern Hemisphere features a number of constel-lations that are prominent in the winter's active night skies: Canis Major, Cetus, Eridanus, Gemini, Lynx, Orion, Perseus, and Taurus.

For starters, you can effortlessly spot the well-known

Open Cluster M38

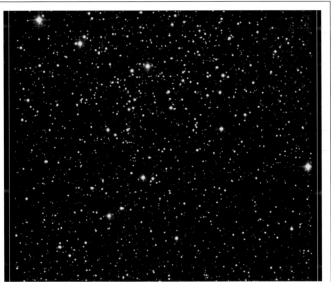

- At 4,200 light-years away in the Auriga constellation is the open cluster M38. This bright cluster of stars can be spotted in the constella-tion's southern region.

- Binoculars and basic telescope viewers some-times detect a cross shape within the multiple stars of the cluster. More powerful telescopes see many more stars and no defined shape to the cluster.

- While in Auriga, check out the famous double star Capella, which is high in the sky in wintertime and ten times bigger than the sun.

Andromeda Galaxy

- The Andromeda galaxy, M31, is located in the constellation that gave it its name. Andromeda is a prominent northern constellation.

- The Andromeda galaxy is considered the most distant body that can be seen with the naked eye. It appears as a weak smudge to the unaided eye, but with binoculars and telescopes, the galaxy becomes more pronounced.

- The Andromeda galaxy has a radiant point of light in the vicinity of its galactic center.

quadrilateral containing three stars in a neat row close to its center. This places your eyes on the constellation Orion the Hunter. The top left star is known as "Betelgeuse," which is pronounced "beetlejuice," by the way. And the bottom right star in the straight line is Rigel. The three stars make up what is called "Orion's Belt." And both Rigel and Betelgeuse are among the top ten brightest stars as seen from Earth.

But it doesn't end there: Right below this belt of stars is the Orion nebula, which exudes a well-dispersed and conspicuously noticeable glow in the crisp and clear winter night sky.

Orion Nebula

- Behind this cavernous region containing blankets of dust and dense gas is the birthplace of multiple thousands of stars.

- Massive young stars choreograph the colors and contours of the Orion nebula, which can be seen most prominently in the Northern Hemisphere's winter.

- The Orion nebula, M42, is 1,500 light-years away from Earth in the constellation Orion the Hunter. It is the closest and brightest star-breeding ground.

Aldebaran in Taurus

- Aldebaran, the brightest star in the constellation Taurus, is a fine wintertime sighting.

- To find Aldebaran, trace Orion's belt from right to left and continue on a vertical line. You cannot miss this orange star that is forty times bigger than our sun.

- Aldebaran is 65 light-years away, which is why—considering its massive size—it looks as small as it does.

- Aldebaran is part of an asterism in Taurus known as the "Bull's Head." Fittingly, it is the bull's eye.

SPRING NIGHT SKIES
The spring night sky is a transitional sky from the more brilliant winter vista

Although decidedly less spectacular than the wintertime and the summertime night skies, springtime night skies supply stargazers with ample celestial fodder. They are not—by any stretch of the imagination—closed for the season. Spring in the Northern Hemisphere is essentially a transitional period for the celestial sphere, with the leading and once highly visible winter constellations slowly but surely exiting stage left and into the farthest recesses of the western horizon before they vanish altogether.

The Northern Hemisphere has constellations of stars that are noticeably high and easily discernible in spring nighttimes, including Bootes, Cancer, Crater, Hydra, Leo, and Virgo. For

Can't Get Enough of the Big Dipper?

- The circumpolar Big Dipper, a bright asterism in the constellation Ursa Major, rests more or less overhead in springtime.

- This asterism has long been known in the night sky and used, too, as a star-hopping point to nearby constellations and stars, including Arcturus in Bootes and Spica in Virgo.

- Native American legend believed that the bowl of the Big Dipper represents a great bear and that the stars forming its handle are warriors chasing after it.

- Various cultures saw in the Big Dipper many things, including a cart and a plow. The Chinese likened the Big Dipper to the government—go figure.

Antennae Galaxies in Corvus

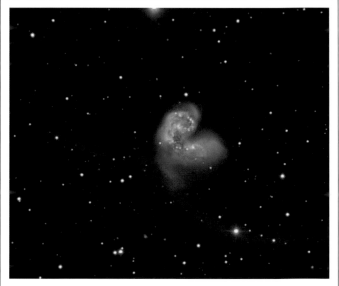

- The Antennae galaxies (NGC 4038, NGC 4039) are in the midst of a galactic collision. They are visible in the constellation Corvus the Crow.

- The galaxies received their names courtesy of the protracted trails of dust, gas, and stars expelled upon their cosmic rear-ender.

- It is believed that approximately 1.2 billion years ago, the Antennae galaxies were individual galaxies: NGC 4038, a spiral galaxy, and NGC 4039, a barred spiral galaxy.

- In four hundred million years or thereabouts, the Antennae's two nuclei will collide.

starters, it's worth checking out the southern horizon to locate Hydra. Although it's not especially bright, you can nonetheless make out a veritable chain of stars that extends southeast to what is known as "Hydra's Head." And although the constellation Cancer is also dim in contrast with so many others, it reveals its contours best during the spring months and houses the M44 open cluster within its borders. This intriguing group of stars is well worth investigating. And also plainly evident at this time of year is the renowned Great Globular cluster, M13, which is a popular target of binocular users. M13 is located in the constellation Hercules the Hero. M13 is a spherical collection of stars that is tightly bound by the forces of gravity.

There are also some highly visible bright stars to track down during the spring season, including Arcturus of Bootes, Vega of Lyra, and Spica, the main star in the constellation Virgo. Virgo is the second-largest constellation of stars. And it's easy enough to locate by tracing the curve of the Big Dipper to Arcturus of Bootes and then venturing from there along the same arc. The especially bright Spica tells you that you've found what you are after.

Spring Starwatching

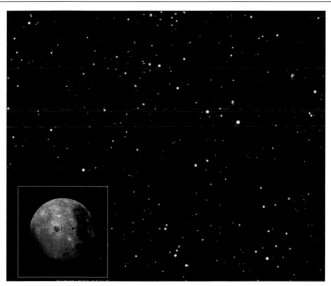

- There are many highly visible stars to check out in springtime, including Arcturus of Bootes, Vega of Lyra, and Spica of Virgo.

- Arcturus, for one, has the distinction of being the most luminescent star in the northern celestial hemisphere and fourth brightest in the entire night sky.

- As is always the case, less moonlight is better for star hunting. Even the very brightest stars are harder to locate with a luminescent moon in the sky.

Great Globular Cluster

- The Great Globular cluster, M13, is nicely visible in both the spring and summertime in the Northern Hemisphere.

- This spherical pool of stars, wound extra tight by the forces of gravity, is quite a celestial sighting in the constellation Hercules.

- This megacollection of stars orbits the Milky Way's galactic core.

- The Great Globular cluster consists of mostly older stars—several hundred thousand of them—and resides in the galaxy's halo.

SUMMER NIGHT SKIES

The summer night skies teem with celestial objects, including the Milky Way's core

After springtime's relatively serene night skies, the Northern Hemisphere's summer season brings with it a brimming plate full of celestial sightings. Among the prominently high summer constellations in the Northern Hemisphere are Aquila, Cygnus, Hercules, Lyra, Ophiuchus, Sagittarius, and Scorpius.

The summer night skies in the Northern Hemisphere are indisputably the most explored. Amateur astronomers, in particular, are out and about at this time of year, starwatching in their own backyards or vacationing in locales that supply impressive picture windows into the busy summer night sky. For starters, check out the famous Summer Triangle of stars, which graces the southeast horizon. This rather considerable

Coma Cluster of Galaxies

- The Coma cluster of galaxies, found in the dim and obscure constellation Coma Berenices, is an intriguing telescope sighting in summertime.

- Hugging the north periphery of Coma Berenices, which is located between constellations Leo and Bootes, is a group of galaxies estimated at some 320 million light-years away.

- The two brightest galaxies in the cluster are elliptical galaxies NGC 4889 and NGC 4874, believed to be twice as wide as the Milky Way.

Summer Triangle of Stars

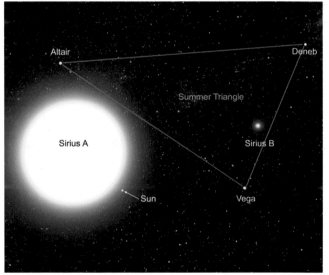

- The Summer Triangle is an asterism containing three bright stars from three constellations.

- This celestial triangle of stars is seen overhead throughout the summer months in the Northern Hemisphere.

- The three stars are all famous in their own right: Vega of Lyra, Deneb of Cygnus, and Altair of Aquila.

- Throughout the months of July through December, Vega—the brightest star of the troika—is practically overhead during the twilight hours and easily spotted.

triangle consists of three bright point stars: Deneb of Cygnus, Vega of Lyra, and Altair of Aquila. The summertime is simultaneously the moment to locate the Great Square of Pegasus, an asterism containing three stars in Pegasus and one in Andromeda.

Perhaps the biggest ticket in the Northern Hemisphere's summer night skies is the Milky Way. That is, you can catch an exceptional glimpse of our parent galaxy's galactic core. Peering into the constellation Sagittarius, you are in essence viewing the plane of the galaxy from its periphery.

Prominent Milky Way

- The Milky Way is a sight for sore eyes in the summer months. Even though it's not as bright as in the Southern Hemisphere, it nonetheless impresses.

- Numerous constellations benefit from the Milky Way's summertime sheen.

- The Milky Way can be seen wending its way behind the Summer Triangle of stars.

- Beginning in Cassiopeia in the northeast sky, our parent galaxy cuts an elongated, bright but nonetheless fuzzy swath across the Summer Triangle to Scorpius in the southwest.

Great Square of Pegasus

- Outside of the Big Dipper, the Great Square of Pegasus is the most recognizable asterism of stars in the Northern Hemisphere night sky.

- The four stars that form this celestial square are actually in two separate constellations: three in Pegasus and one in Andromeda. The stars are Scheat—a reddish giant at its top—and Markab, Algenib, and Alpheratz.

- Alpheratz is located in the Andromeda constellation.

- The Great Square of Pegasus stands out because its interior appears dark.

FALL NIGHT SKIES

Autumn night skies bridge the more active Northern Hemisphere summer and winter night skies

Not unlike the springtime, the Northern Hemisphere's fall night skies are more or less in-between skies. That is, they are not as active and bright as either the summer or winter night skies. Indeed, during the fall months the Summer Triangle of stars can be spotted on the far western horizon just before it disappears from view. And at the same time, many stars associated

with the famous wintertime constellations can be seen rising on the far eastern horizon. This is how the night skies operate all year long. They are perpetually in motion. Autumn nights are good times to appreciate this celestial dynamism.

Nevertheless, the fall night skies in the Northern Hemisphere are not lacking worthwhile sightings in their own

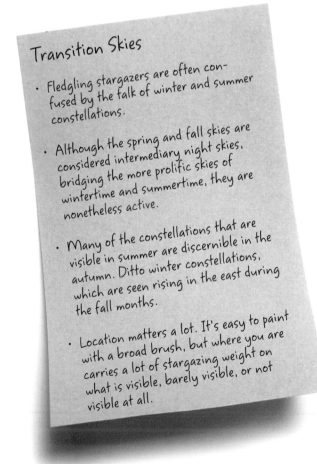

Transition Skies

- Fledgling stargazers are often confused by the talk of winter and summer constellations.

- Although the spring and fall skies are considered intermediary night skies, bridging the more prolific skies of wintertime and summertime, they are nonetheless active.

- Many of the constellations that are visible in summer are discernible in the autumn. Ditto winter constellations, which are seen rising in the east during the fall months.

- Location matters a lot. It's easy to paint with a broad brush, but where you are carries a lot of stargazing weight on what is visible, barely visible, or not visible at all.

Delphinus the Dolphin

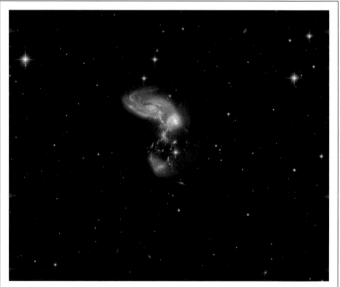

- The fall is a good season to check out some often-overlooked and smaller constellations like Delphinus the Dolphin.

- When you comb the night sky to locate the more obscure constellations, you cannot help but discover others.

- Bordering Delphinus to its north is Vulpecula the Fox. In a clockwise direction, the other neighbors of this constellation are Sagitta the Arrow, Aquila the Eagle, Aquarius the Water Carrier, Equuleus the Little Horse, and Pegasus the Flying Horse.

right. Constellations associated with this time of year include Andromeda, Aquarius, Capricornus, Pegasus, and Pisces. Aquarius the Water Bearer is, in fact, quite high and visible on the southwestern horizon during the fall months. Pisces the Fishes also supplies autumn starwatchers with a view of its Circlet, a petite but nonetheless intriguing loop of stars—that is, the head of its western fish.

Fall is also the season to investigate some of the smaller constellations of stars that don't otherwise get much attention. A fine example is Draco the Dragon, which is observable on the northwest horizon. Other smaller constellations to check out on clear autumn nights include Sagitta the Arrow, Delphinus the Dolphin, and Equuleus the Little Horse. And an unusual and interesting fall nighttime sky sighting is the constellation Eridanus the River. This so-called celestial river of stars is the second-longest constellation. Its leading star, Archernar, can be detected in the southeast night sky as part of the constellation's extended meandering.

Draco the Dragon

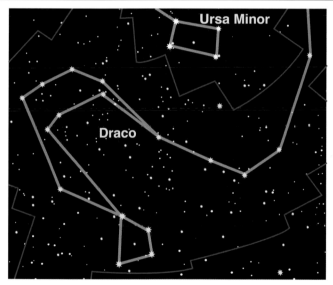

- Because it is a circumpolar constellation, Draco the Dragon is visible throughout the year. Autumn is nonetheless a good time to check it out.

- Revolving around the North Pole as it does, Draco is visible only in Northern Hemisphere skies.

- To find Draco, look for its head, which is a trapezoid found north of Hercules. Its tail of stars meanders along the night sky, ending somewhere between the Big Dipper and Small Dipper.

- At the fringe of the constellation is the star Thuban, the former North Star.

Eridanus the River

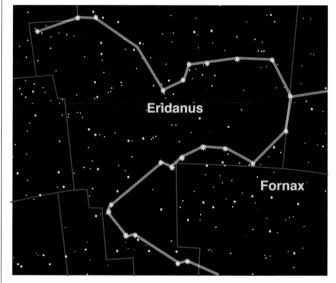

- Eridanus the River is worth seeking out because it is so unique among the constellations. It's known as a celestial river and twists and turns for quite a distance.

- As a stargazer in the Northern Hemisphere, you will spy only the top half of this roundabout constellation.

- Locate it by identifying the star Cursa, which is right next to Rigel in Orion.

- Its brightest star Archernar is found at its most southern tip, which is rarely perceptible in the Northern Hemisphere.

THE SOUTHERN HEMISPHERE

The Southern Hemisphere encompasses everything south of the equator: more oceans and fewer people

The Southern Hemisphere represents the half of planet Earth south of the equator, which is the invented horizontal line that runs completely around the globe at 0° latitude. From a stargazing perspective, the Southern Hemisphere is less well known than its northern counterpart. And this isn't because there is less to see down south—quite the contrary. It's

merely because the Southern Hemisphere has fewer people, more ocean waters, and considerably less landmass than up north. Only 10 percent or so of the world's population resides in the Southern Hemisphere, which includes the continents of Antarctica, Australia, approximately one-third of Africa, and most of South America.

Southern Hemisphere

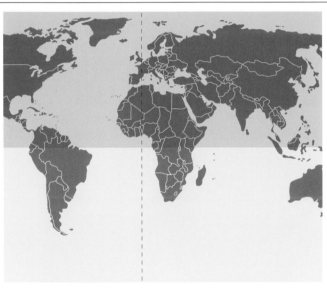

- The Southern Hemisphere encompasses everything that is south of the equator, which includes countries ranging from Australia to New Zealand to Chile to South Africa.

- The Southern Hemisphere's ocean waters dominate its landmass.

- While a small area of South America lies in the Northern Hemisphere, just about everything south of the Amazon River is not.

- The Southern Hemisphere's Antarctica has the highest average elevation of all continents.

Earth's Lower Half

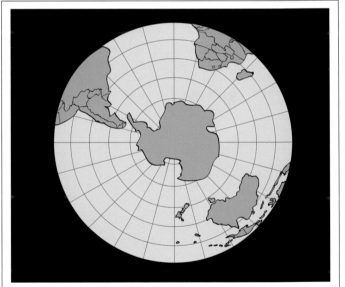

- All of Australia and most of South America lie in the Southern Hemisphere.

- Generally speaking, the Southern Hemisphere sports somewhat milder temperatures than the Northern Hemisphere. This is because it contains more ocean waters than landmass.

- Areas near bodies of water, which tend not to cool down as fast, are typically warmer places.

- This milder exception to the Southern Hemisphere does not include the continent of Antarctica, which is actually a colder locale than its polar opposite to the north.

These lopsided demographic facts are what supply the Northern Hemisphere's night skies with more attention. That is, more astronomy-themed books, star maps, and so forth are oriented to the Northern Hemisphere with its vastly greater numbers of people.

But demographics aside, the Southern Hemisphere's nighttime skies are legendary, renowned for being brighter and more star-laden than their hemispheric opposite. You see, the South Pole in the Southern Hemisphere is actually skewed toward our galaxy's bright galactic core. This positioning is what makes the

Southern Hemisphere night skies decisively brighter than up north, which in turn enables more stars to reveal themselves in the Milky Way's sheen. However, this same heightened brightness detracts from viewing objects in deeper space. Finally, it must be noted that along with the Southern Hemisphere's minimal population comes less pollution. With generally cleaner air, starwatchers get a better view of the night sky up above. Industrialization is the bane of stargazing.

Southern Hemisphere Night Sky

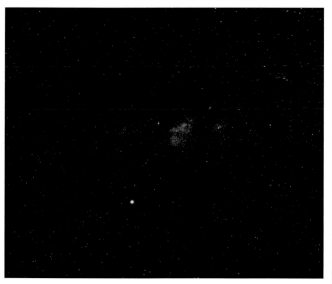

- In the Southern Hemisphere, parts of our galaxy's bright galactic center are overhead in the constellation Sagittarius.

- These celestial coordinates ensure that the Southern Hemisphere's night skies are categorically brighter than the Northern Hemisphere's.

- The Milky Way dramatically reveals itself as an expansive band of muted light arching across the southern night skies.

- The Milky Way's nightlight, if you will, ensures that there are countless more visible stars in Southern Hemisphere night skies.

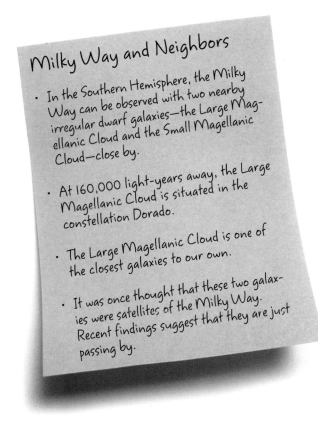

Milky Way and Neighbors

- In the Southern Hemisphere, the Milky Way can be observed with two nearby irregular dwarf galaxies—the Large Magellanic Cloud and the Small Magellanic Cloud—close by.

- At 160,000 light-years away, the Large Magellanic Cloud is situated in the constellation Dorado.

- The Large Magellanic Cloud is one of the closest galaxies to our own.

- It was once thought that these two galaxies were satellites of the Milky Way. Recent findings suggest that they are just passing by.

DISTINCTIVE FEATURES

The Southern Hemisphere night sky is brighter than its Northern Hemisphere counterpart

The Southern Hemisphere has many distinctive features worth pointing out that make stargazing south of the equator a wholly unique and always rewarding experience. Foremost, the Southern Hemisphere's seasons are polar opposites of the Northern Hemisphere's; when it's summer in the north, it's winter in the south.

Most notably, the southern night skies are extremely well lit courtesy of the Milky Way's intense galactic center, which illuminates the night sky in ways never duplicated in the north. This additional luster permits stars that would otherwise be cloaked in faraway shadows to be visible on the celestial sphere. And courtesy of this extra sparkle to the night sky,

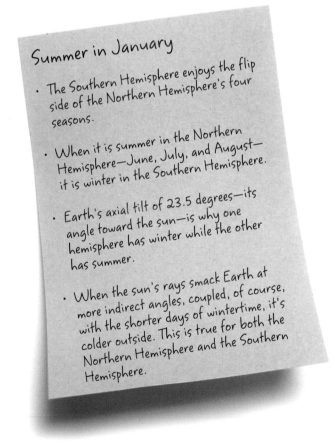

Summer in January

- The Southern Hemisphere enjoys the flip side of the Northern Hemisphere's four seasons.

- When it is summer in the Northern Hemisphere—June, July, and August—it is winter in the Southern Hemisphere.

- Earth's axial tilt of 23.5 degrees—its angle toward the sun—is why one hemisphere has winter while the other has summer.

- When the sun's rays smack Earth at more indirect angles, coupled, of course, with the shorter days of wintertime, it's colder outside. This is true for both the Northern Hemisphere and the Southern Hemisphere.

Upside Down Moon

- Observing the moon from a Southern Hemisphere perch supplies viewers with a distinctly different picture.

- For viewers accustomed to a Northern Hemisphere image of the moon, it appears upside down in areas below the equator.

- With the Southern Hemisphere on the other side of Earth's globe, as it were, it stands to reason that the moon is seen from a wholly different angle.

- Of course, *upside down* is a relative term. Residents of the Southern Hemisphere think otherwise.

certain constellations of stars are considered southern specialties. Canis Major, the Great Dog, and Aquarius the Water Bearer immensely benefit from the Milky Way's glow.

But what Northern Hemisphere observers find most curious about the Southern Hemisphere night skies—those who are lucky enough to sample from both hemispheres—is the apparent inversion of the celestial objects up above. That is, everything appears upside down—from the moon to constellations like Leo the Lion, Bootes the Herdsman, and Orion the Hunter. And so, many of the constellations, which got

their appellations because of perceived shapes resembling gods, animals, and myriad other things, are unrecognizable in Southern Hemisphere night skies. Remember that the majority of constellations of stars received their names from people who studied star groupings and their contours from Northern Hemisphere locations. And, no, when inverted, a lion doesn't look like a lion and a hunter doesn't look like a hunter.

Centaurus the Centaur

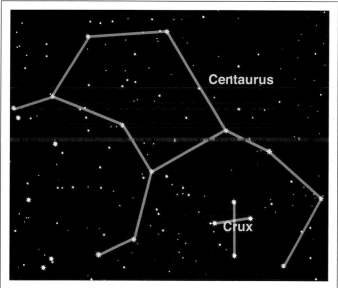

- The constellation Centaurus the Centaur is conspicuously radiant in the southern night skies.

- Centaurus accommodates the Alpha Centauri star system. To the unaided eye, Alpha Centauri appears to be a single star, the brightest star in the constellation.

- Telescopic observation, however, reveals a binary star system. That is, two stars orbiting a shared center of mass.

- Proxima Centauri, a faint red dwarf star, is gravitationally rooted to this system. This is the star nearest to our solar system.

Inverted Hunter

- Constellations appear upside down in the Southern Hemisphere.

- Orion's famous "hunter's belt" of stars doesn't fit the bill from this perspective. Fortunately, most of the constellations' stellar boundary lines do not much resemble the things they are supposed to look like. Orion the Hunter is one of the few exceptions.

WINTER NIGHT SKIES
Southern Hemisphere winter skies supply a potpourri of celestial sightings from stars to nebulae

Winter nights in the Southern Hemisphere deliver a fine show all around. During this season, with the Milky Way providing a well-lit backdrop, a considerable band of constellations graces the night skies. Of course, locating Crux the Southern Cross is job one for stargazers south of the equator. This particular constellation is a hemispheric icon.

Crux, from the Latin term meaning "cross," appears on the southwest horizon during the winter months. Ironically, Crux, along with its "pointers" stars Alpha Centauri and Beta Centauri, is at once one of the most viewed and sought out constellations and the smallest among those officially recognized. The magnificence of Crux the Southern Cross is that

Fomalhaut in Piscis Austrinus

- Fomalhaut, which ranks as the seventeenth-brightest star in the night sky, is found in the obscure southern constellation Piscis Austrinus.

- Piscis Austrinus borders several constellations: Capricornus, Microscopium, Grus, Sculptor, and Aquarius.

- Worth noting is that Fomalhaut was the first star found to have an extrasolar planet, known as "Fomalhaut b," orbiting it. Courtesy of the Hubble space telescope, a visible light image proved what was long suspected.

Crux the Southern Cross

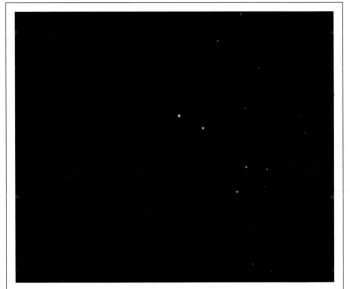

- Crux the Southern Cross is the most renowned constellation in the Southern Hemisphere, even though it is, ironically, the smallest of all eighty-eight constellations.

- Crux is surrounded on three sides by the constellation Centaurus.

- The brightest star in its celebrated cross is Acrux, which is actually two stars orbiting one another.

- Because the south celestial pole is devoid of a bright star akin to the North Star, seafaring explorers long used the cross for guidance.

it literally appears cross-shaped in the night sky, although some argue that it's more kite-shaped. Nevertheless, its distinctive shape sets it apart.

Very near Crux in the busy Southern Hemisphere winter sky is the Coal Sack, an exceptionally dark nebula that reveals itself as a shadowy patch in the elongated Milky Way. The Coal Sack is actually a dense cloud of dust and gases that conceals millions of stars behind it. It is plainly discernible with the naked eye during the winter.

Coal Sack in Crux

Hercules Globular Cluster

- During wintertime, a dark nebula known as the "Coal Sack" is particularly visible—even to naked eye—in the southeastern corner of the constellation Crux.

- It appears as a not-inconsiderable dark patch along the sprawling Milky Way, distinguishing itself among the otherwise illuminated winter sky.

- Despite its clouds of mysterious dark matter, the Coal Sack is, in fact, a fertile star field.

- Portions of the Coal Sack, which reveal some dim light breaking through.

- Although the Hercules globular cluster, M13, is a famous Northern Hemisphere deep-sky object, it nicely reveals itself in the Southern Hemisphere's winter skies.

- Located in the constellation Hercules, the cluster houses more than 300,000 stars— mostly old and bright stars.

- It has been said that this globular cluster is chock full of ancient suns.

- The Milky Way's halo is replete with globular clusters full of aging stars like the Hercules globular cluster.

SPRING NIGHT SKIES

Springtime skies in the Southern Hemisphere remain bathed in the Milky Way's extensive sheen

Spring night skies in the Southern Hemisphere are teeming with activity. The Milky Way's intense core, which is visible as a bulge of yellowish light, is a leading feature. Its durable brilliance dominates the surrounding space. Constellations worth checking out include Sagittarius the Archer, Pavo the Peacock, Equuleus the Little Horse, and Delphinus the Dolphin.

Within the sinuous southern segment of the bright Milky Way is Triangulum Australe, the Southern Triangle, which is easily observed at this time of year. This small constellation is a Southern Hemisphere favorite. Springtime also offers views of both the Small Magellanic Cloud and Large Magellanic Cloud, two neighboring galaxies to the Milky Way.

Pleiades Star Cluster

Triangulum Australe

- Springtime in the Southern Hemisphere is as good a time as any to hunt down both the nearest and brightest of its kind: the Pleiades star cluster, M45.

- This relatively young bastion of hot blue stars is located in the constellation Taurus.

- The cluster can be viewed with the unaided eye, but binoculars kick up the viewing pleasure a couple of notches.

- Bluish reflector nebulae sheath some of the brighter clusters of stars.

- Although the constellation Triangulum Australe, the Southern Triangle, doesn't cover a lot of outer space, it is pored over an awful lot.

- The Southern Triangle itself consists of three stars that form something akin to an equilateral triangle in the celestial ether.

- The stars are the orange-colored Alpha TrA (upper left), Beta TrA (upper right), and Gamma (lower right). Alpha TrA is also known as "Atria."

- The Milky Way's copious dust clouds are highly evident in the vicinity.

On the northern horizon is the Great Square of Pegasus, a considerable square of stars. The constellation Andromeda, which accommodates the Andromeda galaxy within its confines, is also plainly evident. And not to be missed in the spring months is the Pleiades star cluster—the closest of its kind to Earth—with its exceptionally incandescent and hot blue stars. To the east, the night skies supply you with a first-class view of Rigel, a bright star in the Orion constellation.

Regal Rigel

- Spring in the Southern Hemisphere is an interesting time to locate the bright star Rigel in Orion.

- Only the upper part of the constellation will be visible on the horizon, but you will notice a section of Eridanus the River wending its way right past it.

- Rigel is a bright supergiant star with a companion. This secondary star is often lost in Rigel's brilliance.

- The pale glow of the Orion nebula is also visible in the vicinity of Rigel.

Making the Most of the Off-season

- Not everything is viewable from everywhere at all times of the year.

- However, many constellations and their stellar inhabitants are around for the better part of the year.

- Granted, they are often located in less accessible regions of the night sky.

- This is precisely why the stargazing off-seasons of spring and fall are ideal times to hunt down objects that you might have overlooked during primetime, which for most people are the summer months.

SUMMER NIGHT SKIES

The night skies in the Southern Hemisphere's "dog days of summer" are full of activity

Summertime supplies Southern Hemisphere night-sky watchers with a wealth of fine sightings, including the always-visible Large Magellanic Cloud and Small Magellanic Cloud. These irregular dwarf galaxies appear as almost appendages of the Milky Way. And they are circumpolar, which means they are always observable in southern skies.

But although they can be spotted all year around, the summer months furnish starwatchers with keen views of both of them. In fact, the Large Magellanic Cloud is so vibrant in the night sky that even a full moon cannot obscure it.

Summer is a fine time of year to locate famous stars. In the east, you'll encounter Sirius the Dog Star in Canis Major. Sirius

Milky Way Appendage

Small Magellanic Cloud

- The Large Magellanic Cloud, so named in honor of explorer Ferdinand Magellan, is an irregular galaxy with no defined structure.

- This deep-sky object is a Southern Hemisphere gem—a sprawling and diffuse cloud of gas and dust.

- The LMC flaunts only one-tenth of the mass of the Milky Way. It nonetheless accommodates active star-forming regions that are apparent in the celestial miasma.

- The LMC is situated on the border of constellations Dorado and Mensa.

- The Small Magellanic Cloud is a premium summer view in the Southern Hemisphere.

- With the Large Magellanic Cloud due east, it is a perceptibly foggy patch of light in an active summer night sky.

- The Small Magellanic Cloud, an irregular dwarf galaxy, is located in the Tucana the Toucan constellation.

- It is both a member of the Local group of galaxies and one of the nearest to Earth.

is the brightest star in the entire sky. Other stars to look for are Canopus in Carina and Capella in Auriga. Capella is the sixth-brightest star. A pair of stars highly visible in the summer months can be found, too, in Gemini. Castor and Pollux, the "heavenly twins," are responsible for the famous zodiac constellation's moniker.

The summer months in the Southern Hemisphere night skies also open nice portals into constellations such as Carina, Orion, Gemini, Cetus, and Hydra. Within the Hydra constellation is the Hydra globular cluster, a fine summer night's sighting.

ZOOM

The expression "dog days of summer," which we associate with the hot and humid heart of summertime, dates back to ancient Greece. In those days of yore—from early July through early September—the ultrabright Sirius the Dog Star appeared in the sky immediately before or during sunrise. These summer days were considered wicked times and not so happily dubbed the "dog days."

Hydra Globular Cluster

Canopus Supergiant

- Summer in the Southern Hemisphere is a fine time to check out the Hydra globular cluster, M68.

- This cluster of stars, 33,000 light-years away, is located in the constellation Hydra the Water Snake.

- Hydra is the most sprawling constellation in the entire sky, just edging out Virgo for the honor. Along the southern horizon, it wends its way quite a fair distance from Libra to Canis Minor.

- It takes this considerable constellation better than six hours to completely rise in the sky.

- In the southern constellation Carina, Canopus appears white to naked-eye observation.

- Some 310 light-years away, Canopus is a supergiant star believed to be fifteen thousand times brighter than our sun.

- Canopus has been crowned the most inherently luminous of all stars within 700 light-years of Earth.

- Sirius the Dog Star is the brightest star in the night sky. But it is not nearly as bright as Canopus. The fact that Sirius is only 8.6 light-years away wins the day.

FALL NIGHT SKIES

Fall night skies brim with celestial objects, including nebulae and star clusters

The autumn night skies in the Southern Hemisphere are categorically more prolific than their autumn equivalent in the north. A mother lode of compelling sightings is at your disposal, including diverse nebulae and eye-catching clusters of stars. The Milky Way's sheen continues, too, to aid and abet stargazers during this season of the year.

Astronomer E. E. Bernard dubbed the Scutum star cloud the "gem of the Milky Way." Positioned in a strawberry-shaped quarter of our parent galaxy in the constellation Scutum, the Scutum star cloud is indeed worthy of the praise. No undue space dust impedes the view of this radiant cloud of stars. And autumn is a fine time to explore its countless celestial

Pavo the Peacock

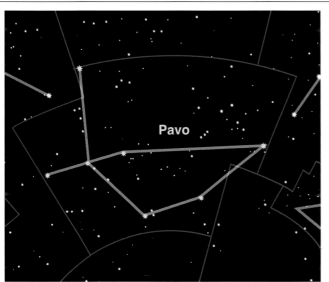

- Pavo the Peacock is a modern constellation.

- The constellation's most noteworthy star, Delta Pavonis, is remarkably sun-like in both its dimensions and mass. It, however, is on its way to becoming a bona fide red giant star. That is, Delta Pavonis has run dry

- of its hydrogen fuel and is, alas, nearing its end.

- Delta Pavonis is 19.9 light-years away from Earth.

- The constellation Pavo the Peacock has been used as a guidepost for locating southern terrain.

Omega Centauri Globular Cluster

- Visible to southern stargazers, the Omega Centauri globular cluster, NGC 5139, is an enormous orb of stars—more than ten million of them—in the constellation Centaurus.

- This cluster is the brightest and most sprawling of its kind.

- Omega Centauri's star population is significantly older than our sun. There are varying ages in the mix, but some of the stars are believed to be twelve billion years old.

- This globular cluster is 150 light-years in diameter.

inhabitants. Ditto the Omega Centauri globular cluster, which wears the crown as the largest of its kind in the Milky Way. The Omega Centauri globular cluster is located in the constellation Centaurus. It is believed to house multiple millions of stars. So many stars, in fact, that estimates place the average distance between them at 0.1 light-years. That may be an unfathomable distance to us on Earth, but it's a stellar stone's throw. And courtesy of its awesome brightness, the Omega Centauri globular cluster is readily seen with the naked eye.

The fall night skies continue unabated to showcase the Milky Way's entrenched constellations, which greatly benefit from the galaxy's warm embrace. With its center still clearly visible in the constellation Sagittarius, there are many additional constellations to survey, including the less explored like Pavo the Peacock, Octans the Octant, and Apus the Bird of Paradise. The western sky showcases additional obscure constellations like Corvus the Crow, Crater the Cup, and Sextans the Sextant. Leo the Lion is nicely positioned in the autumn months with its two brightest stars—Regulus and Denebola—distinguishing themselves.

Apus the Bird of Paradise

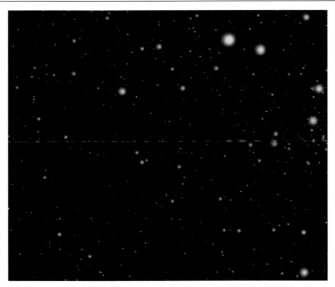

- Although faint in the night sky, Apus the Bird of Paradise is worth checking out in the Southern Hemisphere's autumn sky.

- Apus is one among fifteen circumpolar constellations.

- It can be found between the southern celestial pole and the well-known Southern Triangle in Triangulum Australe.

- Look closely in Apus for three bright stars that form what resembles a small triangle.

Scutum Star Cloud

- The Scutum star cloud is located in the small constellation Scutum the Shield.

- Astronomers have described the cloud as strawberry-shaped.

- It is situated in an area of the Milky Way where little interstellar dust obscures its overall brightness.

- Near the midsection of the Scutum star cloud is the Wild Duck cluster, M11, a grouping of stars that may or may not, depending on your perspective, resemble ducks in flight.

PHASES

The moon's ever-changing angles vis-à-vis the sun and Earth supply us with perpetual phases

Paradoxically, moonlit nights are unwanted intruders into general starwatching endeavors. This is because Earth's only natural satellite casts varying amounts of reflected sunlight into the dark night sky. And an amateur astronomer's best friend is darkness.

However, in its own right, the moon has long been a popular target of night-sky aficionados. If the moon is your astronomical quarry, its well-lit persona works to your benefit. As virtually everybody knows, the moon graces the night skies in never-ending series of phases, which last almost an entire calendar month and then repeat the process. The various phases of the moon are the consequence of its continuously

Lunar Cycle

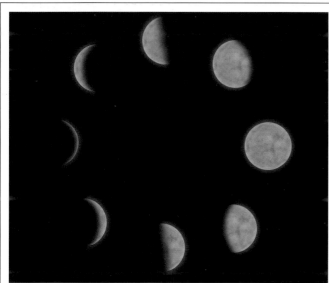

- The moon experiences eight phases during each lunar cycle: new moon, waxing crescent, first quarter, waxing gibbous, full moon, waning gibbous, last quarter, and waning crescent.

- To fully grasp the moon's various phases, acquaint yourself with four words: *crescent, gibbous, waxing,* and *waning.*

- *Crescent* refers to less than half illumination; *gibbous,* more than half illumination.

- *Waxing* implies increasing illumination; *waning,* decreasing illumination.

New Moon Out of the Shadows

- Except during daytime solar eclipses, the new moon is not discernible.

- This phase finds the moon in relative alignment with Earth and the sun.

- The moon is situated between Earth and the sun, and only its far side, which we never see, is illuminated.

- As the moon continues its orbit around Earth, it finds itself aligned again, in about two weeks, with the sun and Earth. Earth, and not the moon, is in the middle of this go-around.

shifting angles as it revolves around Earth. The moon's position vis-à-vis both Earth and the sun is what illuminates, and sometimes obscures, parts or all of it.

The first phase—the new moon—occurs when the sun, moon, and Earth are in relatively close alignment with the moon in between and revealing only its dark, nonilluminated side. During this opening salvo, the moon is completely imperceptible to the naked eye.

The first quarter and last quarter phases—or "half moons"—occur when the moon is situated at a 90-degree angle in relation to the sun and Earth. Simply understood, the phases unfurl like this: After the indistinguishable debut of the new moon, its sunlit segment expands. While the moon is less than 50 percent illuminated, it is said to be in a waxing crescent phase. After it is past 50 percent illumination, the moon is waxing gibbous. On the other side of the full moon, with its sunlit portion now in decline, the moon has entered a waning gibbous phase. Past its second half-moon marker, the waning crescent takes over until it disappears from view altogether.

Half Moon

- The first quarter moon is frequently referred to as a "half moon."

- During this phase, we observe the moon in the night sky exactly half lit—not one-quarter illuminated.

- But because we cannot see its far side, we are, in fact, viewing one-quarter of the moon illuminated by sunlight.

- The half moon occurs when the moon is positioned at a 90-degree angle vis-à-vis the sun and Earth.

Full Moon

- When the moon is aligned with Earth and the sun but not sandwiched between them, we see a full moon.

- During a full moon, the moon is located on the far side of our planet.

- We are treated this time around to the moon's sunlit side.

- The full moon can be appreciated the night before its peak when it's approximately 97 to 99 percent illuminated. Even two days after peak time, while officially in its waning stage, the moon is still 93 to 97 percent illuminated.

IDEAL OBSERVATION TIMES

The prime time to explore the moon's surface is not during its full moon phase

From a dedicated moonwatcher's perspective, the best time to observe the moon is not—contrary to what most people surmise—during its full phase. Sure, when you peer up into the night sky at a full moon, it is an impressive sight, particularly when it has a compelling earthly stage such as an ocean, mountaintop, or city skyline to play on.

But if you are interested in surveying the moon's intricate surface with a pair of binoculars or a telescope, the general consensus is that the full moon is not an optimal observational moment. This is because a full moon is incredibly luminescent. So bright, in fact, that it appears almost flat—a one-dimensional Post-it in the night sky. And, really, looking at the

KNACK NIGHT SKY

Full Moon Looming Large

- Although the full moon is a sight for sore eyes in the night sky, it is extraordinarily bright.

- Looking through binoculars or telescopes at a full moon can be more of a blinding experience than a moon-gazing home run.

- This glare makes observing the moon's distinctive and intricate features difficult. Preferred observational times are between the new moon and full moon phases.

Waxing Gibbous Moon

- The waxing gibbous moon is a fine time to explore its surface.

- During this phase, you can clearly decipher the terminator. Close to the terminator is where you glean the sharpest contrast in surface features.

- The prior moon phases, waxing crescent and first quarter, are also more favorable observation times than a full moon.

- Although it's called a "half moon," the first quarter moon is significantly less than half as bright as a full moon.

full moon through a telescope can be downright painful to the eyes. The sheer brightness supercedes everything else.

Granted, there are no absolute best moon observational times. Nevertheless, there is widespread agreement that several days into the first quarter phase furnish you with prime views of the lunar environment. Both quarter moon phases showcase sunlight emanating from either side of the moon, which permits sunrises and sunsets—from our earthly perspective—to cut a swath right through the moon's heart. Such indirect lighting permits the moon's features to shine.

GREEN ● LIGHT

If you want to survey the intricacies of the lunar landscape, plot out your astronomical adventures when there's a noticeable "terminator" slicing across the moon, which is most prominent in its quarter phases. The terminator is the line that separates night and day on the moon. During a full moon, there is no visible terminator. With a discernible terminator, lunar maria and lunar highlands are cast into marked relief.

Harvest Moon

- The harvest moon, as it's popularly known, is the full moon nearest to the autumnal equinox.

- Sometimes the harvest moon, which is closer to the horizon, appears orange, the result of reflected light passing through more layers of Earth's atmosphere.

- During autumn, there is no measurable lapse between the sunset and moonrise. Farmers were said to have had benefited from the early moonlight as they brought in their bountiful fall harvest.

- The harvest moon is sometimes referred to as a "hunter's moon."

Crescent Moon

- The crescent moon phase embodies a visual moon at less than 50 percent illumination.

- The lack of excessive glare during this phase allows for various night-sky couplings during the year, like the moon and the planet Mars.

- Venus is also seen in the night sky along with the moon at various points in time.

- Because of its unique curvature, the crescent moon is a favorite observational target.

EARTHSHINE

Some of the moon's finest visuals come when it is wholly discernible courtesy of earthshine

When you gaze into the early night sky—not too long after sunset—and chance upon a crescent moon, you might spy something curious. If you're fortunate enough, you'll witness not only the normally vivid sunlit sliver of the moon but also its darkened section, ordinarily invisible to the naked eye. You'll be able to discern the moon's complete contours in dim light. This phenomenon is known as "earthshine." For moonwatchers, earthshine fashions an atmospheric—some might even say "mysterious"—aura in the night skies.

What, in fact, you are witnessing with earthshine is the sun's light reflecting off our planet's surface to the moon's surface, which then casts the light right back at us.

Earthshine at Moonrise

- Earthshine is light reflected from Earth that is visible on the moon's night side.

- Earthshine is evident just before or just after the new moon phase. That is, when the moon is either waxing or waning crescent.

- Courtesy of earthshine, the moon's otherwise-darkened side dimly glows.

- Earthshine permits you to see the moon's entire orb during its crescent phases.

Lunar Perspective of Earthshine

- Earthshine occurs in close proximity to the imperceptible new moon phases because, contrarily, Earth is fully illuminated from the perspective of a lunar horizon.

- That is, sunlight from our planet's radiant glow is reflected back to—from our vantage point, at least—the mostly darkened moon.

- During a waxing crescent moon, earthshine can be seen immediately after sunset.

- During a waning crescent moon, earthshine occurs in the early morning hours.

To put all of this in some stargazing perspective: When the sun's light bathes the surface of the moon, we here on Earth bear witness to its various phases as it orbits our planet. But when the same sun's light that warms and sustains life on Earth is reflected to its only natural satellite, we are the beneficiaries of earthshine. From our vantage point on Earth, this patently diluted light supplies an intriguing contrast with the moon's crescent shapes. Earthshine furnishes stargazers with captivating images of the moon's full orb, even during these partial phases.

THE MOON

Lovely Couple: Moon and Venus

- Because the moon is in its crescent phases when earthshine occurs, it is sometimes observed simultaneously with the planet Venus.

- Venus forms a compelling celestial duet with a moon bathed in earthshine's soft light.

- Because earthshine manifests itself either just after sunset or in the early morning hours, Venus and a moon basking in earthshine are not an uncommon visual.

- The waxing crescent moon has been referred to as the "old moon in the new moon's arms."

Early Evening Crescent Moon

- An early evening waxing crescent moon rising above treetops is prime time for witnessing earthshine.

- As the moon rises higher in the sky, earthshine's effects dissipate.

- Scientists have discovered that other planets in our solar system display their version of earthshine. Planetshine, as it's called, has been seen on some of Saturn's moons.

- Even Saturn's expansive and luminous rings have showcased a version of earthshine onto the planet proper.

LUNAR MARIA

The dark, flatter areas of the moon visible from Earth are the lunar maria

When ancient astronomers looked up at the moon in the night sky, they detected seemingly flat, shadowy areas that noticeably contrasted with its more rugged—and more abundant—lighter sections. They assumed that these darker shades of the moon were extensive bodies of water, or oceans, and named them *maria* or the singular *mare*—the

Latin term for "seas." And although these stargazers from yesteryear were off beam in their guesswork, the moniker stuck.

The lunar maria are volcanic plains that comprise approximately 17 percent of the moon's entire surface. Courtesy of radiometric dating, it is believed that this rocky terrain is the byproduct of volcanic eruptions that occurred 3.16 to

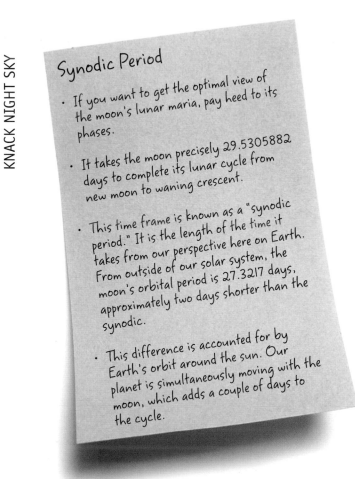

Synodic Period

- If you want to get the optimal view of the moon's lunar maria, pay heed to its phases.

- It takes the moon precisely 29.5305882 days to complete its lunar cycle from new moon to waning crescent.

- This time frame is known as a "synodic period." It is the length of the time it takes from our perspective here on Earth. From outside of our solar system, the moon's orbital period is 27.3217 days, approximately two days shorter than the synodic.

- This difference is accounted for by Earth's orbit around the sun. Our planet is simultaneously moving with the moon, which adds a couple of days to the cycle.

Lunar Maria on the Moon

Mare Imbrium
Oceanus Procellarum
Mare Serenitatis
Mare Crisium
Mare Insularum
Mare Vaporum
Mare Tranquillitatis
Mare Cognitum
Mare Humorum
Mare Fecunditatis
Mare Nubium
Mare Nectaris

- Approximately 17 percent of the moon's surface is covered with sprawling, shadowy balsitic plains called the "lunar maria."

- The various regions of the moon with lunar maria are thus named for seas, such as Mare Vaporum (Sea of

Vapors), Mare Humorum (Sea of Moisture), and Mare Frigoris (Sea of Cold).

- With their iron-intensive compositions, lunar maria are poor reflectors of sunlight—hence their dark appearance.

4.2 billion years ago. *Apollo 11* disembarked on a small lunar mare called the "Mare Tranquillitatis," otherwise known as the "Sea of Tranquility."

The various lunar maria throughout the moon's surface have been supplied names to differentiate them from one another. Names include Mare Imbrium (Sea of Rains), Mare Nectaris (Sea of Fertility), and Mare Nubium (Sea of Clouds). Lunar maria rocks are iron-rich basaltic lavas not unlike Hawaiian volcanic debris.

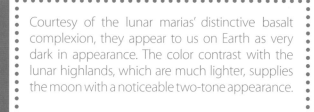

ZOOM

Courtesy of the lunar marias' distinctive basalt complexion, they appear to us on Earth as very dark in appearance. The color contrast with the lunar highlands, which are much lighter, supplies the moon with a noticeable two-tone appearance.

Mare Tranquillitatis

- The Mare Tranquillitatis (the Sea of Tranquility) is the most famous of all the lunar maria.

- It is the landing site of the *Apollo 11* lunar module.

- Three small craters to the north of the landing site have been named "Armstrong," "Aldrin," and "Collins" in the *Apollo 11* astronauts' honor.

- In contrast to other maria, the Mare Tranquillitatis displays a lightly bluish tint, which is likely the byproduct of higher metal content in its rocky terrain.

Mare Nubium

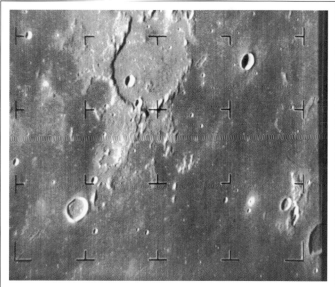

- The moon's Mare Nubium, the "Sea of Clouds," reveals numerous meteorite craters subsequently filled with flowing lava. Igneous rock—basalt—has formed over time as the lava solidified.

- Basalt has been found not only on the moon but also on the planets Mars and Venus, indicating past volcanic activity there.

- When culled from the lunar maria, so-called moon-rock basalts—rich in iron and titanium—are very different from rocks collected in the more numerous highlands.

LUNAR HIGHLANDS

The majority of the moon's surface consists of old and heavily cratered highlands

In contrast to the dark, relatively flat terrain that defines the lunar maria, the remainder of the moon's surface—which happens to be its majority at near 83 percent—consists of lunar highlands or terrae. The highlands are at odds with the maria, which can be observed in their decidedly contrasting hues, with the highland areas appearing light and the maria dark.

The lunar highlands received their fitting tag because of their mountainous appearance and elevation, which set them apart from the level maria. There are, in fact, numerous well-known highlands that are situated alongside the edges of huge impact basins crammed with mare basalt.

The lunar highlands, though, are not basaltic but rather

Two-Tone Moon

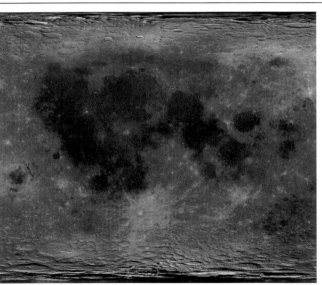

- The lighter areas visible on the moon are known as the "lunar highlands."

- They are the higher altitude regions of the moon and have many more impact craters than the lower-altitude maria.

- The moon's far side consists of predominantly highland topography and few maria.

- The highlands consist of different rock than are found in the maria. They are covered with anorthosite, a rock consisting of lighter elements like calcium and aluminum. This is why the highlands are lighter in appearance.

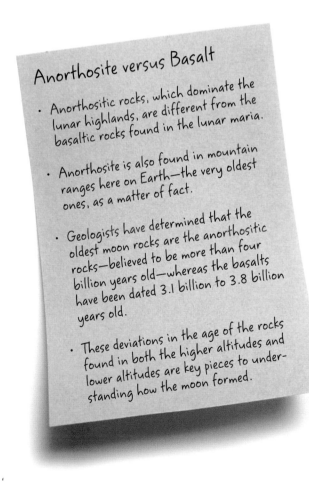

Anorthosite versus Basalt

- Anorthositic rocks, which dominate the lunar highlands, are different from the basaltic rocks found in the lunar maria.

- Anorthosite is also found in mountain ranges here on Earth—the very oldest ones, as a matter of fact.

- Geologists have determined that the oldest moon rocks are the anorthositic rocks—believed to be more than four billion years old—whereas the basalts have been dated 3.1 billion to 3.8 billion years old.

- These deviations in the age of the rocks found in both the higher altitudes and lower altitudes are key pieces to understanding how the moon formed.

anorthositic in makeup. The scientific consensus is that the highlands came to be when feldspar crystallized and worked its way to the pinnacle of the oceans of molten lava generated by volcanic activity during the moon's formation.

Although both the lunar maria and lunar highlands are cratered, the highlands are significantly more wracked with the residue of outer-space wreckage. Without Earth's protective atmosphere, the moon has been the recipient of countless meteoroid and asteroid strikes throughout its existence of more than four billion years.

Interestingly, the far side of the moon, which we can never see from our vantage point on Earth, is almost entirely highlands, with few maria on the surface. Hence, it is light colored and a veritable monolith when compared with the side that faces Earth. What we see when looking at the moon are visibly contrasting shades, which are what supplies it with a seeming countenance—that is, the man in the moon.

Southern Highlands

- The moon's southern highlands reveal a multifaceted surface. That is, numerous overlapping craters.

- This particular area of the moon—with its densely cratered mien—furnishes observers with a picture window into its dramatic desolation.

- The craters upon craters found in the southern highlands supply scientists with a look at both the old and the new. The craters that have superimposed themselves on others are obviously a more recent vintage.

Apollo 16 Landing Site

- The *Apollo 16* was the fifth mission to the moon in which men walked the surface and returned to Earth with samples of their cosmic adventure.

- This mission—from April 16 to April 27, 1972—has the distinction of being the first during which a module touched down in the moon's mountainous highland region.

- *Apollo 16* landed in an area of the moon known as the "Descartes Highlands."

IMPACT CRATERS

Without a protective atmosphere, space objects have continuously bombarded the moon, producing countless craters

If there is one feature that is synonymous with the moon, it is craters. The moon is scarred by literally millions of them—big and small. They range in size from the minuscule found in individual rock pieces to over 200 miles (360 kilometers) in diameter. There are over forty so-called impact basins on the surface of the moon. Craters achieve this ballyhooed distinction based on having rim diameters in excess of 185 miles (300 kilometers).

For centuries a scientific debate raged as to the cause of these many craters dappling the moon's remote and forbidding surface. But after the *Apollo 11* mission and moon landing, with samples of moon rock and soil brought back to Earth, the mystery was largely solved. The craters were the consequence of

Moon's Cratered Surface

- The moon's surface is strewn with millions of impact craters that have fashioned a unique soil called its regolith. With no protective atmosphere to fend off these celestial interlopers, ground-up rock makes for quite a top soil.

- While long thought otherwise, recent findings suggest there may be some water on the moon.

- *Apollo* astronauts left visible footprints on the Moon's surface, indicating the presence of a loose, albeit rocky, soil atop the cratered landscape.

Goclenius Crater

- The moon's Goclenius crater is situated near the outskirts of Mare Fecunditatis.

- Its crater edges are conspicuously worn and asymmetrical in appearance.

- The floor of Goclenius exhibits a protracted, twisted, and narrow channel. This is indicative of past lava flows.

- Rudolph Goclenius (1572–1621) is the man after whom the Goclenius lunar crater was named. A German physician and professor of mathematics, he published *Urania*, a title devoted to astronomy when such books were rare birds indeed.

direct—and in many instances, violent—impacts from meteoroids, asteroids, comets, and other space matter. Here on Earth, with the planet's protective layers of atmosphere, such disfiguring collisions were largely avoided, with potentially debilitating space objects burning up before reaching the surface.

The moon, however, has no such atmosphere. Nor does it contain wind or water to erode the many craters. In other words, the moon's craters remain largely as they were upon initial impact. That is, until another impact comes along and reconfigures the landscape.

Copernicus Crater

- The Copernicus lunar crater is one of the most frequently observed by amateur astronomers. It is situated just northwest of the moon's center.

- This crater is easily spotted with a pair of binoculars or, better yet, a small telescope.

- The Copernicus crater is believed to be only eight hundred million years old—a mere youth as these things go.

- Because of its relative young age, the crater—in contrast with older craters—reveals no evidence of having been flooded with molten lava.

Moon Rocks

- Plain and simply, moon rocks are rocks picked up on the moon's surface.

- Moon rocks brought back to Earth are subject to what is known as "radiometric dating."

- The basaltic samples, which were gathered from the moon's low-lying lunar maria, have been found to be younger than the rocks brought back from the highland areas.

- The ages of dated moon rocks range from 3.1 billion years to 4.5 billion years. The latter number is close to the presumed age of Earth.

MERCURY & VENUS

These are the only two planets closer to the sun than Earth is

The two planets closest to the sun are Mercury and then Venus. Earth is next in the procession. Both Mercury and Venus—along with Earth and Mars—are classified as the terrestrial or innermost planets. The terrestrial planets are all rocky in composition but quite varied in size, atmosphere, and temperature.

Although the planet Mercury is nearest the sun, it is also the smallest of the eight planets in our solar system. Mercury is about 40 percent of Earth's overall diameter. Because of its close proximity to the sun and relatively small size, Mercury is not easily spotted in the night sky. Whereas Earth is some 93 million miles away from the sun, Mercury is, by comparison, 36 million miles away. It orbits the sun approximately once every fifty-nine Earth days. Mercury is dense and has a heavily cratered terrain not unlike our moon's surface.

Venus is sometimes referred to as "Earth's twin" because

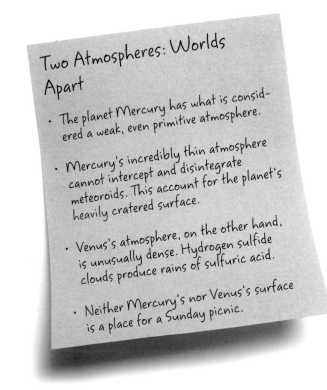

Two Atmospheres: Worlds Apart

- The planet Mercury has what is considered a weak, even primitive atmosphere.

- Mercury's incredibly thin atmosphere cannot intercept and disintegrate meteoroids. This account for the planet's heavily cratered surface.

- Venus's atmosphere, on the other hand, is unusually dense. Hydrogen sulfide clouds produce rains of sulfuric acid.

- Neither Mercury's nor Venus's surface is a place for a Sunday picnic.

Venetian Surface

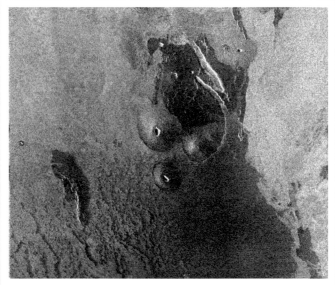

- Venus's surface is visibly shaped by considerable volcanic activity.

- There are also few craters on the Venetian surface. And the highly visible craters are quite large.

- The tremendous atmospheric pressure on Venus does not permit anything resembling life as we define it. No bodies of water could exist under such conditions.

- Venus's air is only ten times less dense than water. In contrast, Earth's air is one thousand times less dense than water.

Earth and Venus are close in size. But there the similarities end. Venus is exceptionally warm and dry. There is no water on the planet because of its scorching temperatures. In fact, getting a good handle on Venus's surface conditions has been problematic because dense clouds of sulfuric acid envelop the planet. Nonetheless, we know that the surface features plains, mountains, canyons, and valleys.

From our perch on Earth, Venus is frequently the brightest object in the night sky, including both stars and planets. Venus can be glimpsed even during daylight. When the planet is nearing Earth during its orbit, it is detectable during the early evening hours. When Venus is moving away from Earth, it is perceptible in the early morning hours. Hence, its nicknames as the "Evening Star" and "Morning Star."

Terrestrial Planets

- Earth is the largest of the terrestrial planets.

- And although Venus is sometimes referred to as "Earth's twin," its surface conditions couldn't be more different. If you were to stand on the surface of Venus, the pressure exerted on you would be ninety times or greater than what you experience here on Earth.

- Yet, despite these horrific surface conditions, at 31 miles (50 kilometers) to 40 miles (65 kilometers) above this horror show, Venus's temperatures and atmospheric pressure are strikingly similar to Earth's.

Runt of the Planet Family

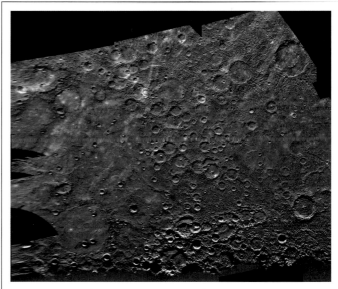

- Mercury's surface is akin to the moon's: heavily cratered and barren in appearance.

- It orbits the sun faster than any of the other seven planets.

- Mercury's average distance from the sun during its orbital path is 36 million miles (58 million kilometers). Earth, on the other hand, is approximately 93 million miles (150 million kilometers) away from the sun.

- The diameter of the smallest planet, Mercury, is only 3,032 miles (4,879 kilometers).

MARS
This always-intriguing planet is visible in the night sky as a reddish star

Mars is one of the four terrestrial or inner planets. It is the fourth planet from the sun following Mercury, Venus, and Earth. From earthlings' perspectives, Mars is the most beguiling and most discussed planet. It is the one planet in our solar system that conceivably could be visited by a peopled spacecraft someday.

Mars is sometimes referred to as the "Red Planet" because of

its pinkish-red shades visible to astronomical observers. This color scheme is the byproduct of widespread iron oxide on the planet's surface. Mars's atmosphere is thin and clear. Clouds can thus be spotted now and again, as can occasional dust storms caused by solar heating of the razor-thin atmosphere and the subsequent rapid movement of air, which kicks up

Red Planet

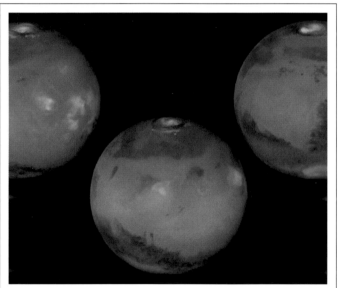

- Mars has a diverse surface that includes soaring volcanoes, far-reaching valleys, stretches of desert, and polar ice caps.

- High concentrations of iron oxide on the surface are what scientists believe give the planet its reddish tint in the night sky.

- Mars's deserts are noticeably rust-colored.

- Mars's southern hemisphere looks much like our moon. However, its northern hemisphere, with few craters, is conspicuously different in appearance.

Martian Surface

- Mars's surface showcases a volcano, the Olympus Mons, which dwarfs Hawaii's Mauna Loa.

- A sprawling valley in Mars's Tharsis region, Valles Marineris, spans 2,500 miles (4,000 kilometers) and is, at its widest point, 120 miles (220 kilometers).

- At Valles Marineris's deepest point of 4 miles (5 kilometers), its depth surpasses fourfold that of the Grand Canyon.

- This Martian valley covers close to two-thirds of the planet's diameter.

incredible amounts of dust. At their most violent, these storms are renowned for entirely enveloping the planet.

Mars is a fascinating planet on a wide range of fronts, including its hosting of the Olympus Mons, the tallest mountain in our solar system. Mars has familiar attributes to the denizens of its sister planet, Earth: valleys, deserts, volcanic terrain, and even polar ice caps. But it also sports elements common to the moon, such as large impact craters that stretch for miles. The Borealis Basin, as it's called, spans nearly 40 percent of the planet's surface.

Mars's Bigger Moon

- Phobos is the bigger of Mars's two moons. It is also closer to its parent planet than Deimos—the second moon—and thus more photographed and studied.

- Phobos is peculiarly shaped. To call it "nonspherical" is an understatement.

- Phobos orbits nearer to its parent planet than any other known moon.

- It was speculated for a spell that the bizarrely shaped Phobos might be hollow. It's not!

The Face of Mars

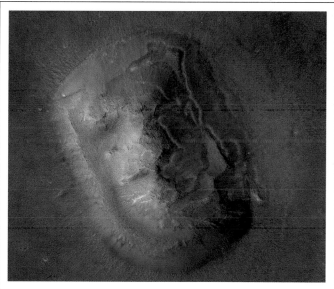

- The Cydonia region of Mars is perhaps its most famous area. Popular culture has seen to that by playing up the so-called face of Mars, which is visible in the Cydonian hills.

- Indeed, one of the hills resembles a human face from certain angles.

- This area in Mars's northern hemisphere is accessible to telescopic viewings.

- Recently more detailed images from space probes reveal, alas, a rather pedestrian Martian hill.

JUPITER

Gaseous Jupiter is by far the largest planet in our solar system

A popular target of amateur astronomers, the planet Jupiter bears the distinction of being the fourth-brightest celestial object. Only the sun, moon, and Venus exhibit more luminescence for the benefit of starwatchers. Jupiter is the fifth planet from the sun and the first so-called outermost planet. In stark contrast to the rocky terrestrial planets, which include Earth, Jupiter is composed of mostly gases: chiefly hydrogen and helium.

Jupiter is frequently referred to as a "gas giant" because of both its gaseous nature and humongous size. In fact, this one planet touts two and one-half times the combined mass of the seven other planets in our solar system. It has 318 times the mass of Earth. But because thick and intricate clouds consisting of ammonia crystals continuously shroud Jupiter, probing this behemoth is not so easy. That is, Jupiter's core innards are still not fully understood.

Atmosphere and Surface

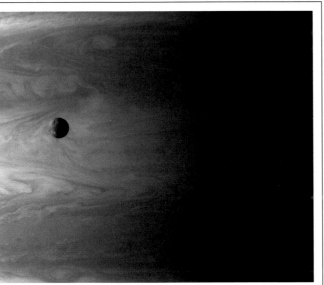

- Jupiter's atmosphere is its surface.

- What you are seeing when eyeing Jupiter are the tops of dense clouds high up in the planet's atmosphere.

- Jupiter has no solid surface to speak of, but its gaseous innards nonetheless get denser as they get closer to the planet's core.

- Like its gas giant cousin, Saturn, Jupiter has rings around it, but they are considerably smaller and fainter. Even powerful telescopes have a hard time deciphering Jupiter's rings.

Great Red Spot

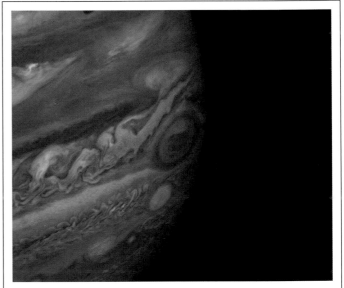

- Bands of the oval-shaped Great Red Spot on Jupiter are visible to even small telescopes.

- The Great Red Spot consists of gaseous cloud tops both appreciably higher and colder than neighboring regions of the planet's atmosphere. The Great Red Spot is large enough to accommodate two Earths.

- Although Venus is the brightest object—after the moon—in night sky, it is nonetheless rarely seen in the late evening hours. At this point, Jupiter is the brightest "star."

You see, Jupiter's atmosphere runs extremely deep. So deep that it's quite possible the atmosphere is the entire planet. Jupiter could be a planet consisting of entirely gas, sans any rock-based or firm surface like Earth and the other terrestrial planets. One of the spectacular visual spectacles surrounding Jupiter is its Great Red Spot and other spots, the result of the planet's eternal clouds and complex weather events repeatedly altering its appearance.

Galilean Moons

- With a good pair of binoculars, Jupiter's four Galilean moons are visible in the night sky. Jupiter has fifty-seven lesser-known moons.

- Were it not gravitationally bound to Jupiter, Ganymede would be a planet in its own right. And Callisto is practically Mercury's dop-

pelganger in size and look.

- Europa is considered the smoothest natural object in our solar system. Contrarily, Io, with its active volcanoes spewing sulfuric acid, looks like a celestial pizza pie floating in space.

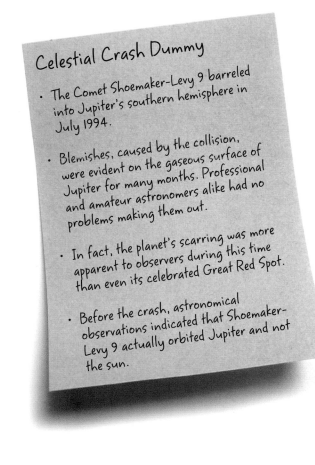

Celestial Crash Dummy

- The Comet Shoemaker-Levy 9 barreled into Jupiter's southern hemisphere in July 1994.

- Blemishes, caused by the collision, were evident on the gaseous surface of Jupiter for many months. Professional and amateur astronomers alike had no problems making them out.

- In fact, the planet's scarring was more apparent to observers during this time than even its celebrated Great Red Spot.

- Before the crash, astronomical observations indicated that Shoemaker-Levy 9 actually orbited Jupiter and not the sun.

SATURN

Courtesy of its luminous rings, Saturn is one of the most recognizable solar system bodies

The image of the planet Saturn is familiar to people of all ages in all earthly locales. Courtesy of its lustrous ring structure, it is widely appreciated as an object of sheer beauty in our solar system. It is, therefore, of great interest to stargazers. In fact, Saturn was the outermost planet known to ancient observers of the night sky. Uranus was originally thought to

be a star, and Neptune was not discovered until more powerful telescopes made it possible.

Saturn is the second-largest planet in overall dimensions. Like Jupiter, it is classified as a "gas giant" because of its distinctive composition. Indeed, the planet with the impressive ringlets is an enormous orb of gases with no evidence of a

Saturn Is a Gas

- Just like Jupiter, the planet Saturn is entirely made up of gases with no solid surface.

- Thick bands of gaseous clouds surround the planet.

- Scientists believe that Saturn may, in fact, have a solid nucleus consisting of

iron and rocky matter.

- Although Saturn is massive—sporting ninety-five times the mass of our home planet—it is the least dense of all the planets. It is only one-tenth as dense as Earth.

Saturn Close-Up

- The stunning rings of Saturn cannot be distinguished with small telescopes.

- Flyby satellites like Cassini-Huygens supply us with the most intimate looks at this extraordinary planet.

- What we see as Saturn's surface are thick clouds of gas.

They blanket the planet and give it a uniform, almost bland look. But sometimes Saturn exhibits cloud patterns akin to Jupiter's Great Red Spot.

- But when all is said and done, it's the rings around its gaseous orb that make Saturn stand apart.

solid surface. Thick cloud layers enshroud it. In other words, what we know as life on Earth could not possibly exist on Saturn. Nevertheless, the planet is thought to maintain a solid, but searing, inner core made of metallic and rocky matter.

Intriguingly, Saturn rotates on its imaginary axis at a speedy clip; only Jupiter spins faster. Saturn takes just ten hours and thirty-nine minutes to complete one rotation; Earth, on the other hand, takes a full twenty-four hours—our definition of a day. Worth noting, too, is that Saturn has many moons orbiting it, including Titan, which is larger than the planet Mercury.

Saturn's Largest Moon

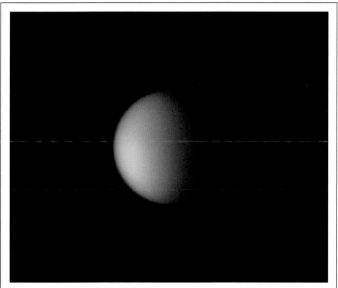

- Titan's diameter measures 3,200 miles (5,150 kilometers), making it larger than the planet Mercury.

- The opaque atmosphere of this sizeable moon gives it an almost uniform yellow appearance to flyby satellite observation.

- Titan's atmosphere consists of mostly nitrogen. Few moons in our solar system have atmospheres as dense as Titan's.

- With its dense atmosphere and the discovery of liquid hydrocarbon lakes in its polar regions, Titan has been compared with early Earth.

Brilliant Rings

- Consisting of ice, dust, and miscellaneous rock chunks, the brilliant rings that surround Saturn do not actually interact with the planet's gaseous surface.

- They are extraordinarily long but thin. They range up to 180,000 miles (300,000 kilometers) across

but are not more than .6 mile (1 kilometer) thick.

- A gap estimated at close to 2,000 miles (3,200 kilometers) or more separates the distinctive rings.

- Saturn sports four major groups and three lesser groups of rings.

URANUS

Uranus is the farthest planet from Earth that can be observed with the naked eye

Uranus is the most distant planet in our solar system that can be seen without the aid of a telescope. Ancient astronomers noted its existence in the night sky but did not classify it as a planet because of its faint illumination and seemingly deliberate orbit. They considered it just another star—one among many in the night sky.

Nevertheless, the night-sky observers from yesteryear were on the mark regarding Uranus and its uneven orbiting patterns. In a class by itself in this regard, Uranus spins on its axis on the same plane as its orbit around the sun. This unusual positioning and movement supply the planet with unpredictable seasons and weather conditions far removed from our

Frozen Planet

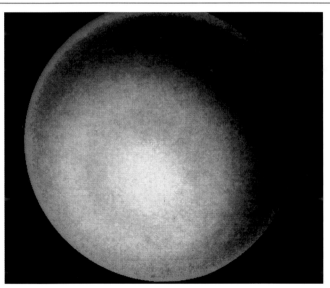

- Uranus is an enormous orb of gases and liquids.

- Its pale blue-green clouds are its defining feature. The planet's omnipresent clouds are composed of minuscule methane gas crystals.

- Although Uranus has no explicit solid surface, it is nonetheless very different from the gas giant planets of Jupiter and Saturn.

- In contrast to the interiors of Jupiter and Saturn, Uranus's interior consists of frozen gases and rock. Its atmosphere's temperature is -355°F (-215° C)—cold indeed.

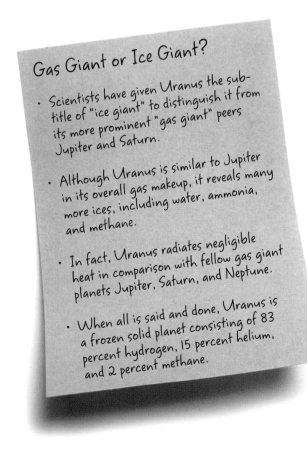

Gas Giant or Ice Giant?

- Scientists have given Uranus the subtitle of "ice giant" to distinguish it from its more prominent "gas giant" peers Jupiter and Saturn.

- Although Uranus is similar to Jupiter in its overall gas makeup, it reveals many more ices, including water, ammonia, and methane.

- In fact, Uranus radiates negligible heat in comparison with fellow gas giant planets Jupiter, Saturn, and Neptune.

- When all is said and done, Uranus is a frozen solid planet consisting of 83 percent hydrogen, 15 percent helium, and 2 percent methane.

winter, spring, summer, and fall climates. A defining Uranus feature is, in fact, its extremely high wind velocities, believed to be as much as 450 miles (725 kilometers) per hour.

Uranus is the seventh planet from the sun and often referred to as an "ice giant," as opposed to "gas giant." Its atmosphere is the coldest among the planets in our solar system. Although the composition of Uranus's atmosphere is similar to the atmospheres on Jupiter and Saturn—mostly hydrogen and helium—its dense cloud layers paint a vastly different picture of its overall makeup.

A fascinating glimpse of Uranus came our way courtesy of NASA's *Voyager 2*, which photographed the planet up close. What this space probe witnessed was a uniform-looking planet sans any distinguishing features—a solid blue-green orb seemingly hanging in space. This lack of character is believed to be photographic confirmation of Uranus's celebrated blue-green clouds consisting of minuscule methane crystals.

Uranus, like its outer-planet brethren, exhibits no distinct surface—no solid outer core. It sports thirteen Saturn-like rings, a magnetosphere, and numerous moons.

Moons of Uranus

- Uranus's major moons are Puck, Miranda, Ariel, Umbriel, Titania, and Oberon.

- Titania is at once Uranus's largest moon and the eighth-most massive in our solar system. It is bigger than Earth—twenty times more massive, as a matter of fact.

- Uranus's moons were all named after characters in works from Alexander Pope and William Shakespeare. Examples include Cordelia from *King Lear*, Ophelia from *Hamlet*, and Desdemona from *Othello*.

Rings Around Uranus

Uranus ▪ *Hubble Space Telescope* ACS/HRC WFPC2

2003   2005   2007

NASA, ESA, and M. Showalter (SETI Institute) ▪ STScI-PRC07-32

- Uranus has a planetary ring system. Although the rings are not as prominent as Saturn's, they are nonetheless highly complex. The planet's outermost Epsilon ring is made up of large ice boulders.

- Uranus's myriad rings are not quite circular and vary appreciably in brightness.

- The planet's inner rings are gray, almost opaque.

- Based on evaluations of the their composition, scientists surmise that Uranus's rings are just mere youths—less than six hundred million years old.

NEPTUNE

Neptune is the planet farthest from the sun and invisible to the naked eye

The planet Neptune has the distinction of being the only planet that cannot be observed by the naked eye at any time. Indeed, a telescope is in order to locate Neptune in the night sky.

Neptune is the eighth—and last—planet from the sun. Pluto, which has been demoted to dwarf planet, is a greater distance away, and it, too, cannot be seen without the aid of a telescope. Neptune is, in fact, thirty times farther from the sun than is our home planet. And whereas Earth orbits the sun once every year, Neptune takes close to 165 years to complete this journey.

Neptune is a diverse planet with eleven known satellites and more than a few visible rings surrounding it. Of its

Most Distant Planet

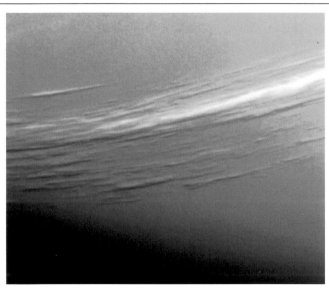

- The rather consistent bluish appearance of Neptune is derivative of the planet's frozen methane clouds, which absorb other colors. Methane gas is the chief chemical found in natural gas.

- Neptune, the eighth planet from the sun, is similar to Uranus in composition.

- Scientists suspect that Neptune houses an Earth-sized core composed of rock-solid matter.

- If Neptune was a hollow shell, and not a gaseous mélange, it could accommodate close to sixty Earths.

Dynamic Neptune

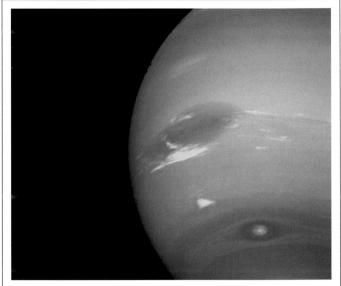

- *Voyager 2* first revealed that Neptune is not a wholly blue monolith but rather a dynamic planet with a lot happening both at its surface and beneath its dense layers of clouds.

- *Voyager 2* informed us that this distant planet is a windy place, too, sporting the fastest winds in our solar system. Winds of 1,200 miles (2,000 kilometers) per hour on Neptune are commonplace.

- *Voyager 2* also confirmed the planet's ring system, which theretofore was only a scientific supposition.

multiple moons, Triton is the largest. Neptune is probably best known for its distinctive blue clouds, which consist of frozen methane gas. These dense clouds are continually in rapid motion, with wind speeds swirling them around at 700 miles (1,100 kilometers) per hour.

Because Neptune has an atmosphere composed of hydrogen, helium, water, and silicates—the latter being the same minerals found in Earth's rock-laden crust—one might surmise that Neptune, too, sports a solid surface beneath its clouds and atmosphere. No such evidence exists.

Neptune's Largest Moon

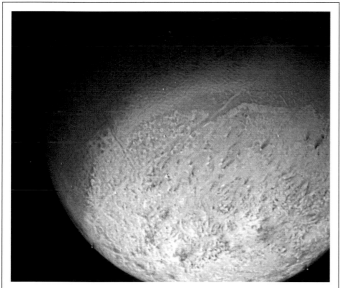

- Triton is Neptune's largest moon of the thirteen known to exist. The *Voyager 2* mission discovered most of the planet's satellites in 1989.

- Triton is frigid. In fact, it's colder than all other bodies measured in our solar system. Its surface temperature registers -391°F (-235° C).

- Triton is also the only significant moon in our solar system that orbits in a counterdirection to its parent planet's orbit.

- Unlike Triton, most of Neptune's moons are small. It is believed that they are mostly former asteroids.

Great Dark Spot

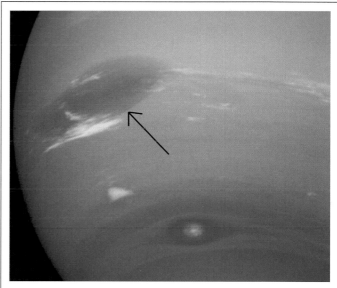

- Neptune's surface sports a Great Dark Spot, not too different from Jupiter's Great Red Spot.

- The Great Dark Spot is located in the planet's southern hemisphere. It is caused by anticyclonic storms. Near the spot, winds have been measured at 1,500 miles (2,400 kilometers) per hour.

- The Great Dark Spot reveals itself in darker shades of blue.

- Unlike Jupiter's Great Red Spot, which lasts decades, the Great Dark Spot comes and goes every few years.

WHAT ARE STARS?

Those twinkling lights in the sky are actually fiery balls of hydrogen and helium gases

To put it in the most understandable vernacular, the countless stars we see in the night sky are flaming orbs of gases: mostly burning hydrogen and helium. The closest star to Earth is, of course, the sun. The sun is blisteringly hot and in the midst of ongoing thermonuclear reactions.

Substantial masses are what keep the sun and its family of

stars from completely detonating and coming apart during the furious nuclear reactions occurring in their interiors. Basically, stars embody such incredible amounts of matter that it takes multiple billions of years for their ongoing gaseous interactions to expend the fuel behind the epic burning.

Astronomers surmise that stars are the natural byproduct of

Starry, Starry Night

- A star-filled sky paints a striking celestial mosaic.

- But stargazing is made even more extraordinary when you fully appreciate what you're witnessing.

- Most of the visible stars in the night sky are, in fact, in the process of converting

hydrogen into helium. And because of their massiveness, it will take billions of years for them to expend their life fuel.

- Stars are unceasingly releasing energy and electromagnetic radiation into the interstellar medium.

The Sum of Their Parts

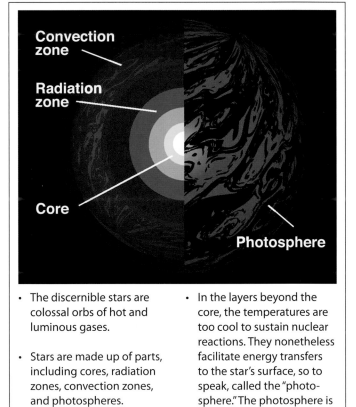

Convection zone
Radiation zone
Core
Photosphere

- The discernible stars are colossal orbs of hot and luminous gases.

- Stars are made up of parts, including cores, radiation zones, convection zones, and photospheres.

- The core, or nucleus, is where the nuclear fusion occurs.

- In the layers beyond the core, the temperatures are too cool to sustain nuclear reactions. They nonetheless facilitate energy transfers to the star's surface, so to speak, called the "photosphere." The photosphere is what we see.

the universe's formation. That is, they believe that the fledgling universe was not much different in composition than an average star is today: extremely hot, dense, and made up of mostly hydrogen, helium, and miscellaneous elements. At the moment of the Big Bang, the widely accepted theory of our universe's beginning approximately 13.7 billion years ago, nuclear fusion on a large scale converted hydrogen to helium—akin to the inner workings of stars—in the more or less familiar ratio of 3:1.

However, beyond the astronomical minutia of stars forming in gargantuan gas clouds and continually discharging energy—electromagnetic radiation—into the celestial ether, these twinkling space objects continue to rivet observers of the night sky. Stars release light generated from their blazing insides. Planets, on the other hand, emit no light whatsoever. They merely reflect light from stars. Our solar system's planets reflect light from the sun, which they all orbit.

The universe is a work-in-progress, with new stars constantly forming and old stars petering out and disbanding as they use up the last of their energy sources, which will be—alas—the fate of our sun in several billion years.

Sagittarius Star Cloud

- The Sagittarius star cloud, M24, is 600 light-years wide. It's a multihued picture window into the diverse star life residing near the center of the Milky Way.

- Most of the stars in the cloud are orange and red in color.

- An individual star's color is reflective of its surface temperature. The blue and green hues indicate the hottest stars in the hodge-podge, signifying younger and more massive stars.

- The bright red-colored stars, which distinguish themselves from the pack, are older red giants.

Night-Sky Panorama

- A sky panorama supplies you with a peek into the diverse family of stars.

- The spectral classification of stars based on their temperatures is blue-violet, blue-white, white, yellow-white, yellow, orange, and orange-red. The hottest stars are blue-violet; the coolest, orange-red.

- Mintaka is a blue-violet star; the sun, a yellow star.

- The high temperature of a star doesn't equate to luminosity: The bright Betelgeuse is a cool super-giant star.

RED GIANT STARS

When stars grow old and weak, their radiation wanes, and they appear red in color

Stars have finite life spans. And, yes, that includes the sun that nurtures and sustains life on Earth. Because stars are systematically in the process of converting hydrogen to helium, they engender incredible amounts of light and radiation, which splash out into the interstellar medium and race through space. The fact that we have observed stars that are billions of light-years away is evidence of their extraordinary illumination. In other words, their light expulsion travels through space billions of years before reaching us here on Earth.

Still, natural and colossal changes are inevitable with nuclear fusion at play. In other words, nothing lasts forever during such intense burning processes, including massive

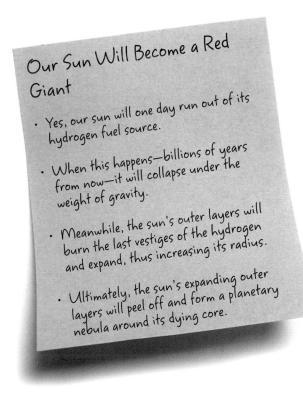

Our Sun Will Become a Red Giant

- Yes, our sun will one day run out of its hydrogen fuel source.

- When this happens—billions of years from now—it will collapse under the weight of gravity.

- Meanwhile, the sun's outer layers will burn the last vestiges of the hydrogen and expand, thus increasing its radius.

- Ultimately, the sun's expanding outer layers will peel off and form a planetary nebula around its dying core.

Betelgeuse in Orion

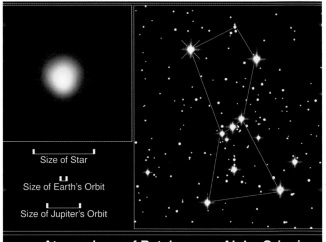

Size of Star

Size of Earth's Orbit

Size of Jupiter's Orbit

Atmosphere of Betelgeuse · Alpha Orionis
Hubble Space Telescope · Faint Object Camera

- Betelgeuse, located in the constellation Orion the Hunter, is a red giant star—one of the largest visible stars in the night sky.

- To find it, cast your eyes or telescope toward the mighty hunter's right shoulder.

- Astronomers have long been trying to assess the literal luminosity of Betelgeuse, which is considered both a red giant and a semivariable star. Estimates place its brightness facility at 85,000 to 105,000 times that of our sun.

- Betelgeuse is 640 light-years away.

stars with billions of birthdays under their belts.

Through time, with the ongoing conversion of hydrogen to helium taking place, the weightier element—helium—sinks to the star's core and assumes a bigger and more defining role in its life and times. The more substantial helium now finds itself surrounded by a sheath of hydrogen. At this point, the ever-decreasing amounts of hydrogen are incapable of manufacturing sufficient energy to keep the star fully intact. The star's death rattle begins with the outermost layers collapsing. Keep in mind that this timetable is not sudden or compressed into a matter of Earth days or even years but rather over billions of years.

At this milestone in a star's life—its last gasp—the hydrogen fusion loses its luster permitting the temperature and pressure to dramatically rise in the star's interior. This happens because burning helium assumes control of the star's fate. Weakened considerably now without its fuel converter doing its thing, the star cools and expands along with its collapsing periphery. These dying stars appear reddish-orange to observers.

Gamma Crucis Above the Coal Sack

- Gamma Crucis, aka "Gacrux," is a red giant star prominently located in Crux the Southern Cross in the vicinity of Coal Sack.

- Its bright reddish and orange hues are easily distinguishable in the night sky. At 88 light-years away, Gacrux is the closest red giant to our planet.

- Gacrux is also a binary star, believed to have a white dwarf companion close by.

- If Gacrux were substituted for our sun, this red giant star's massive sprawl would bring it more than halfway to Earth.

Giants and Dwarfs

- Do not confuse red giant stars with the most common stars found in the universe: red dwarfs. Red dwarfs are small, relatively cool stars with minimal luminosity, which is in stark contrast to red giants.

- The largest red dwarf star observed delivers only 10 percent of the sun's radiance.

- With their smaller masses, cooler temperatures, and slower burning processes, red dwarf stars can likely endure for 10 trillion years.

WHITE DWARF STARS

At the end of some stars' lives, their hot cores carry on as white dwarfs

White dwarf stars are the smallest and faintest stars in the universe. They comprise an estimated 6 percent of all the stars in existence. Fundamentally, what distinguishes these types of stars from their larger and brighter peers is structure.

Although red giant stars are so-named because they are in the final stages of life—and emitting feeble radiation—they nonetheless maintain a measure of nuclear fusion, even if it's in rapid and evident decline. When a star exudes a reddish color, helium, fusing into carbon and oxygen at its core, has assumed the dominant role. At this point, its hydrogen fusion, although demonstrably waning, is still a factor.

A red giant star is red hot, if you will, at its heart, which is

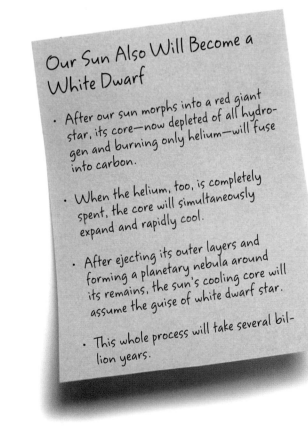

Our Sun Also Will Become a White Dwarf

- After our sun morphs into a red giant star, its core—now depleted of all hydrogen and burning only helium—will fuse into carbon.

- When the helium, too, is completely spent, the core will simultaneously expand and rapidly cool.

- After ejecting its outer layers and forming a planetary nebula around its remains, the sun's cooling core will assume the guise of white dwarf star.

- This whole process will take several billion years.

Dog Star and Companion

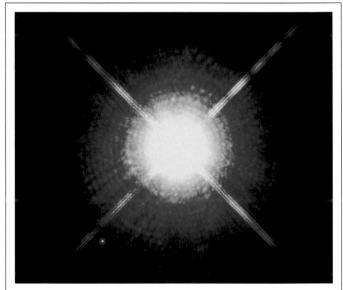

- Sirius the Dog Star is the brightest star. But what most people are unaware of is that Sirius has a white dwarf companion.

- The companion is known as "Sirius B." These companion white dwarfs often go unseen, consumed by the overwhelming brightness of the primary star. But Sirius B is nonetheless there.

- Typically, white dwarf stars are as massive as our sun but not much bigger than Earth.

- This is precisely why they are difficult to discover on the vast celestial frontier.

in stark contrast to its fast cooling and peeling-away outer layers. Red giant stars often spawn planetary nebulae around now. However, when the nuclear fuel of a red giant star is absolutely spent, and its aforementioned burning core is all that remains, a white dwarf star is often the byproduct. That is, the elements remaining in the celestial object, now incapable of generating any sort of fusion, leave the star with no energy source. The faint light that white dwarf stars expel is actually stored thermal energy, which will deplete itself through time.

Globular Cluster M4

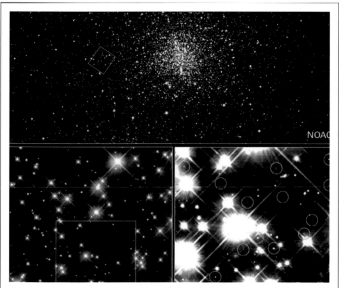

NOAO

- Globular cluster M4, located in the constellation Scorpius, is among the nearest of its kind to Earth.

- Under optimal night-sky conditions M4 can be observed with the naked eye.

- The Hubble space telescope nonetheless supplies us with a close-up view of M4 with its significant population of white dwarf stars.

- These extraordinarily dense stars are quite small. With the naked eye, you will be unable to make out the white dwarfs in the cluster.

Black Dwarfs

- Our sun will be become a red giant and then a white dwarf.

- Scientists also presume that it will, in the final analysis, end its days as a black dwarf.

- When a white dwarf star completely cools off and is no longer generating any sort of energy and commensurate illumination, it is considered a black dwarf.

- But hold on: Black dwarfs are only conjecture. The universe's age is 13.7 billion years, and no star since its formation has lived long enough to become a black dwarf. It takes billions of years for white dwarfs to cool to the temperatures of their cosmic surroundings.

VARIABLE STARS

These stars have changeable brightness due to internal or external factors

A variable star is so-named because of its varying degrees of brightness as detected from Earth. These variations can be the consequence of numerous factors, including changes in a particular star's illumination due to internal disruptions or completely extraneous factors in its outer-space neighborhood.

These variables are divided into two categories: intrinsic and extrinsic. *Intrinsic* relates to specific goings-on within the star's myriad layers and life evolution. This could amount to either enlargement or shrinkage in size, which naturally impacts its luminosity. *Extrinsic* accounts for brightness fluctuations tied to conditions outside the star's body. This could include a nearby celestial object that partially obscures the

The Envelope, Please

- The outermost layers of gas surrounding stars are sometimes referred to as their "envelopes."

- Variable stars often sport rather far-reaching envelopes.

- This extensive gas range means more sparkle emanating from the stars but decreasing density in the envelopes.

- This lessening density in the sprawling layers of gas increases a star's variability period and qualities.

Variable Star Mira

- Mira is a red giant star that is also a variable star. Both its radius and brightness vary.

- Mira, aka "Omicron Ceti," was the first variable star identified as such. Its name has been applied to an entire breed of similar stars called "Mira variables."

- Mira variables pulsate—with the stars' atmospheres apparently expanding and contracting—and are recognized by their strikingly red and orange colors.

- These stars are not very big but they can shine with the brightness of thousands of suns.

star, altering its appearance in the night sky.

Variable stars fascinate night-sky observers. In 1638 when astronomer Johannes Holwarda noticed a particular star pulsating with conspicuously unpredictable illumination, he called this red giant star "Mira," the Latin term for "astonishing," because it behaved like no other previously observed and documented star. Holwarda's variable star find was important because it flew in the face of ancient astronomers and philosophers, like Aristotle, who believed that the night sky is completely staid with no surprises whatsoever.

V838 Monocerotis

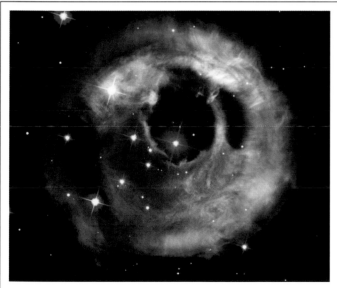

- The mysterious variable star V838 Monocerotis is found in the constellation Monoceros.

- Suddenly, and without warning, this pedestrian star in a little-known constellation glowed with the brightness of 600,000 suns.

- For one brief shining moment, V838 Monocerotis was the brightest star in the entire Milky Way. It has since returned to its former obscurity.

- Astronomers surmise that this blinding outburst was part of an unstable aging star's stellar evolution.

Cepheid Variable Stars

- The Hubble space telescope has located Cepheid variables—massive variable stars—a long way from home.

- These variable stars have been spotted in the NGC 4603 spiral galaxy, 108 million light-years away.

- Cepheid variables are inherently bright but difficult to fully appreciate at such a long distance from Earth.

- These variable stars are renowned for brightening and then dimming from one day to the next.

STARS

CLOSEST STARS TO EARTH

The nearest stars to Earth are not always the brightest in the night sky

The sun is the nearest star to Earth, and, suffice it to say, there is no close second. This trifling detail is an apropos snapshot of our place in the cosmos. The sun and countless other factors all coming together in a meticulous synergy are what make life possible on Earth, as opposed to the seven other inhospitable planets in our solar system. Yet, the neighboring

stars to both the sun and Earth are a long way from the cozy confines of our solar system.

The sun at .000016 light-years away—or 93 million miles—from Earth is a celestial stone's throw when compared with the next-closest stars. After the sun, the nearest star to Earth is Proxima Centauri at 4.2 light-years away, followed by Alpha

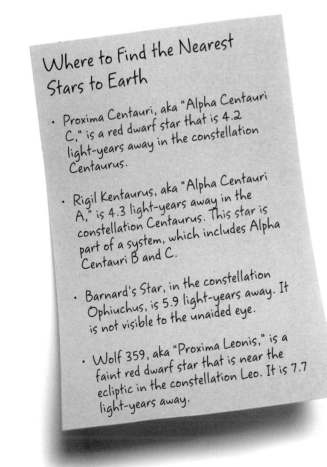

Where to Find the Nearest Stars to Earth

- Proxima Centauri, aka "Alpha Centauri C," is a red dwarf star that is 4.2 light-years away in the constellation Centaurus.

- Rigil Kentaurus, aka "Alpha Centauri A," is 4.3 light-years away in the constellation Centaurus. This star is part of a system, which includes Alpha Centauri B and C.

- Barnard's Star, in the constellation Ophiuchus, is 5.9 light-years away. It is not visible to the unaided eye.

- Wolf 359, aka "Proxima Leonis," is a faint red dwarf star that is near the ecliptic in the constellation Leo. It is 7.7 light-years away.

Nearby Stars

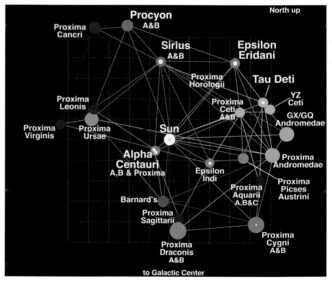

- The nearest stars to Earth, the sun, and our solar system are actually works-in-progress.

- Because the science of astronomy is dynamic and subject to change, you may encounter variations in lists of the closest stars.

- Our stellar neighborhood is nonetheless dominated by red dwarf stars, which are faint and often invisible to the native eye.

- Relatively speaking, these stars are small—only about 40 percent as massive as our sun—and cooler than most of their brethren.

Centauri A and B at 4.3 light-years away. And to further pad this roster of next-door neighbors: Barnard's Star is 5.9 light-years away; Wolf 359 is 7.7 light-years away; Lalande 21185 is 8.3 light-years away; Sirius A and B are both 8.6 light-years away; and A and B Luyten 726-8 are 8.7 light-years away.

But distance aside, one paradoxical facet concerning these, relatively speaking, nearby stars is their overall brightness. That is, from starwatchers' perspectives, the closest stars to Earth are not always the brightest stars in the night sky. The fraternity of stars in the universe is by no means a monolithic collection in size, composition, and—yes—illumination. Among the nearest stars to Earth and, indeed, our solar system, are some that are not visible to the naked eye and can be seen only with a pair of binoculars or a telescope. Most of these stars are red dwarfs, which are too faint to see, from any distance, with the unaided eye.

Proxima Centauri

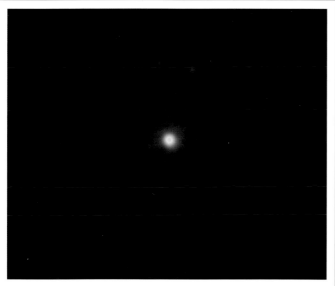

- Proxima Centauri, the nearest star to our solar system, is not anything special as stars go. But the fact that it's the closest star body to our naked eyes, binoculars, or telescopes makes it special indeed.

- As are stellar neighbors Barnard's Star and Wolf 359,

Proxima Centauri is a red dwarf star and faint in the night sky.

- Proxima Centauri is part of the Alpha Centauri system of stars. Proxima Centauri orbits its two companions: Alpha Centauri A and Alpha Centauri B.

Red Dwarf Country

- Red dwarf stars are the most common stars found in our stellar neck of the woods.

- They are at once unremarkable and not very visible stars.

- Red dwarf stars have low surface temperatures in contrast to other types of stars.

- Red dwarf stars nonetheless have long lives. This accounts for their ubiquity. They can thrive for trillions of years—more than the presumed age of the universe.

BRIGHTEST STARS

The brightest stars from Earth's perspective are actually incredibly more luminescent than the sun

The brightest stars in the night sky are not always the closest to Earth. In fact, many of the most luminescent stars are quite a distance away from our planet. Rigel, for example, among the top ten brightest stars from our vantage point here on Earth, is 775 light-years away. Barnard's Star, the fourth-nearest star, excluding the sun, at a mere 5.9 light-years away, cannot be seen without binoculars or a telescope.

Astronomers apply various measuring sticks to assess the overall brightness of stars, including magnitude and luminosity. The latter grades stars' intrinsic brightness. And although nearness to Earth doesn't necessarily mean brighter, distance is a factor in the overall equation. That is, some stars are

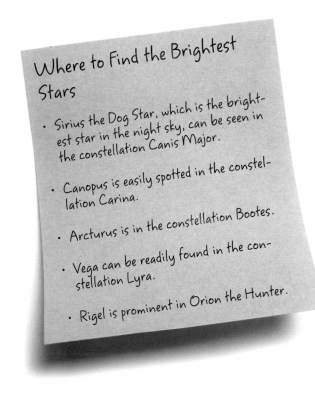

Where to Find the Brightest Stars

- Sirius the Dog Star, which is the brightest star in the night sky, can be seen in the constellation Canis Major.

- Canopus is easily spotted in the constellation Carina.

- Arcturus is in the constellation Bootes.

- Vega can be readily found in the constellation Lyra.

- Rigel is prominent in Orion the Hunter.

Undisputed Champion

- From our perch here on Earth, the sun, of course, is the brightest star bar none. There is no close second.

- Sirius, the most vivid visible star in the night sky, earns its celestial gold medal because it is at once relatively close to Earth and bright in its own right.

- If Sirius replaced our sun, it would shine twenty-three times brighter.

- Meanwhile, nearby Wolf 359, a celestial stone's throw at 7.7 light-years away, is one of the faintest stars known.

extraordinarily bright but are way down on the brightest list because they are so far away. Others are less vivid but relatively close to us and therefore appear brighter in the night sky.

Sirius A, the Dog Star, which got its appellation from the Greek word *seinus*, which means "searing," is the brightest star. With the sun measuring 1 in illumination—the standard that all other stars are measured against—Sirius comes in at 23, which means that it shines 23 times brighter than the sun. At 8.6 light-years away, Sirius is highly visible in the constellation Canis Major the Big Dog.

The bright stars that follow Sirius A are Canopus in the constellation Carina the Keel; Alpha Centauri A and B at the foot of the constellation Centaurus; Arcturus, the brightest star in the constellation Bootes; and Vega in Lyra the Harp.

Ultra-bright Canopus

- Canopus, the second-brightest star in the night sky behind Sirius, is worth tracking down in the constellation Carina. Its extraordinary luminescence is the stuff of legend.

- It is believed that Canopus shines 100,000 times brighter than our sun. Now

that's awfully bright. If you think it's warm in summertime, multiply that by 100,000 and consider the amount of suntan lotion you would need.

- Canopus is visible all year long in the Southern Hemisphere.

Seriously Bright Sirius

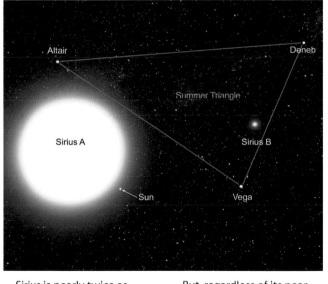

Altair   Deneb

Summer Triangle

Sirius A   Sirius B

Sun   Vega

- Sirius is nearly twice as bright as Canopus, the second-brightest star in the night sky.

- The famous Dog Star is relatively close to our solar system, which naturally enhances its luminosity from our stargazing perspective on Earth.

- But, regardless of its nearness to planet Earth, Sirius is an impressive star.

- Under optimum conditions, Sirius can be seen even in the daylight hours when the sun is low on the horizon.

WHAT ARE FORMER STARS?

Stars end their lives in a variety of fashions, including in spectacular explosions

Nothing lasts forever, not even stars that are billions of years old. Originating out of clouds of interstellar gas known as "nebulae," stars form and live rather diverse lives in a sprawling universe beyond our full comprehension.

Stars with solar masses five times or more than that of our sun are considered gigantic in the big picture. Stars with smaller solar masses—less than five times that of the sun—are more commonplace. These smaller stars are closer in composition and behavior to the sun. And size matters when it comes to a star's death pangs. Both big and small stars' lives eventually come to an end, but they exit the celestial stage in different fashions.

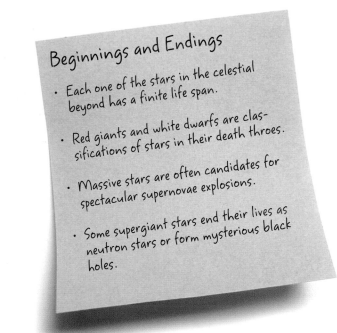

Beginnings and Endings

- Each one of the stars in the celestial beyond has a finite life span.

- Red giants and white dwarfs are classifications of stars in their death throes.

- Massive stars are often candidates for spectacular supernovae explosions.

- Some supergiant stars end their lives as neutron stars or form mysterious black holes.

Planetary Nebula NGC 6543

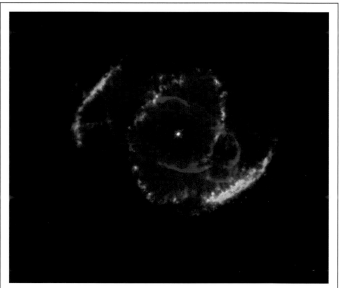

- Planetary nebula NGC 6543, known as the "Cat's Eye nebula," is located in the constellation Draco.

- The center of the Cat's Eye nebula accommodates a very bright and very hot star.

- Planetary nebulae are the consequence of stars in their later stages of life expelling their outer layers into the interstellar medium.

- Even though there is no planet involved, the term *planetary nebulae* is used to describe these gaseous bubbles in space that sometimes encircle dying stars.

Massive stars at the end of their celestial ropes often go out in a blaze of glory, exploding into smithereens and generating extraordinarily colorful and dramatic supernovae. As a robust dead star's remains splatter into the interstellar medium, the debris sometimes forms new stars, planets, moons, and any number of space objects. The aforementioned smaller stars with less mass do not go out in such grand style but instead morph into white dwarf stars. These rather average-sized stars, which burn hydrogen into helium throughout their enduring lives, eventually run low on fuel. This scenario is expected to play out with the sun, which is already several billion years old. The sun's got another several billion or so years of life ahead of it before it begins shedding its outer layers, expanding, and getting brighter in appearance.

Supernova 1987A

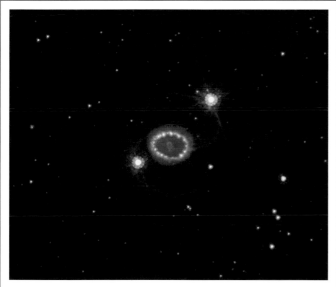

- Supernova SN1987A was spotted in the environs of the Tarantula nebula in the Large Magellanic Cloud.

- This fiery supernova event was actually the nearest to Earth since Johannes Kepler's discovery—sans a telescope—of a supernova in 1604.

- This supernova marked the end of the star Sanduleak's life. Because it is 169,000 light-years away, scientists surmised that this impressive finish to a long stellar life occurred around 167,000 B.C.

- It took 169,000 years for its light show to reveal itself.

Supernova 2007bi

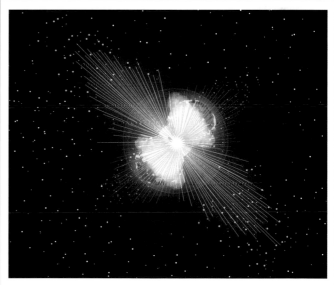

- Supernovae ordinarily emit monumental bursts of radiation that momentarily outshine light from entire galaxies.

- SN2007bi revealed brilliance and staying power that dwarfed those of typical supernovae. Scientists estimated its illumination at fifty to one hundred times brighter than the norm.

- It is believed that the star behind this supernova flaunted a mass of two hundred suns, which would be bigger than any observed or conceived of.

NEUTRON STARS
Neutron stars are sometimes the byproduct of stars bursting into pieces during supernova events

One of the potential afterlives of a former star is what is known as a "neutron star." A neutron star is sometimes a surviving remnant of a dazzling and animated supernova event. Neutron stars are only the offshoots of large stars with solar masses four to eight times more than the sun's. Smaller stars with less mass do not ever spawn neutron stars on their death beds. Our sun,

for instance, will not become a neutron star.

After a massive star depletes its hydrogen supply, which is what kept it hot and bright for billions of years, nuclear fusion is no longer an option. At this point in the star's dwindling life, a supernova explosion of some sort is inevitable. The dying star's outer layers are cast into the celestial ether during the

The Peculiar Neutron Star

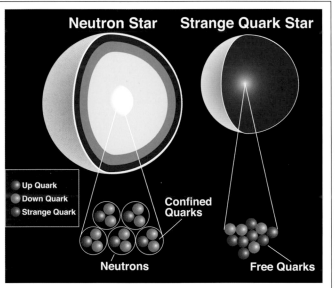

- Neutron stars form rock-solid outer crusts during the gravitational collapses of massive stars at the end of their stellar lives. They nonetheless maintain fluid interiors.

- During supernova explosions, electrons are compelled to fuse with protons, which form neutrons.

- Neutrons frequently stem the total collapse of the star and allow remnants of it to remain intact.

- Strange quark stars made up of theoretical matter are believed to exist when the mass of the neutron star is there but not the requisite dimensions.

Lone Neutron Star

- Neutron stars are practically impossible to detect.

- The Hubble space telescope caught a rare glimpse of one. Scientists concluded they were indeed observing a neutron star because the object was at once extremely hot and extremely small.

- Only a neutron star could be so hot with a diameter of just 16.8 miles (28 kilometers).

- Electrons and protons produce neutrons when compressed. This derivative star matter is therefore dense—unfathomably so, as a matter of fact.

blast as the star's central core collapses under the strain of gravity. However, occasionally during this episode, as protons and electrons join together to form neutrons, an utter collapse of the star is stemmed. When this occurs, a neutron star is a potential remnant—a celestial body that will endure from a star that once was.

The defining attribute of a neutron star is its density. Neutron stars are actually the densest objects known in the universe. Testament to their gravitational sway, there are neutron stars believed to have planets orbiting them.

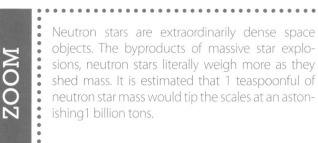

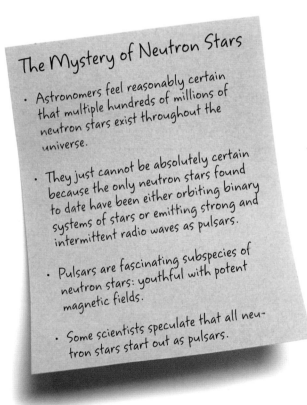

The Mystery of Neutron Stars

- Astronomers feel reasonably certain that multiple hundreds of millions of neutron stars exist throughout the universe.

- They just cannot be absolutely certain because the only neutron stars found to date have been either orbiting binary systems of stars or emitting strong and intermittent radio waves as pulsars.

- Pulsars are fascinating subspecies of neutron stars: youthful with potent magnetic fields.

- Some scientists speculate that all neutron stars start out as pulsars.

Globular Cluster M15

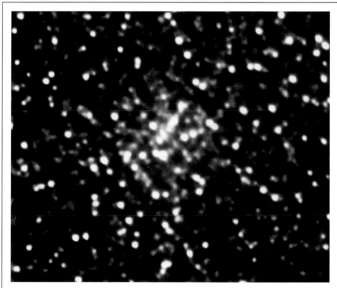

- Globular star cluster M15 accommodates several known neutron stars in its sprawling stellar mix.

- These neutron stars and pulsars are the remains of former supergiant stars, which underwent supernova experiences, from a long time ago.

- It is believed that these neutron stars were born, if you will, when the globular cluster was young and forming.

- M15 is approximately 33,600 light-years away in the constellation Pegasus.

PULSARS
Pulsars are neutron stars that simultaneously spin rapidly and exhibit robust magnetic properties

Pulsars are neutron stars—that is, massive stars reincarnated upon their violent deaths in supernova explosions. They display exceptional qualities that continually fascinate and confound astronomers. What distinguish pulsars from ordinary neutron stars are their rapid rotations—perpetual spinning, as it were—that trigger episodic flashes of radiation.

Like all neutron stars, pulsars are incredibly dense. But they also maintain spirited magnetic fields that spin right along with them. It is believed that pulsars spin at one thousand times or more per second, which generates a gravitational pull of epic proportions. For a little perspective here: It has been said that a marshmallow colliding with the surface of

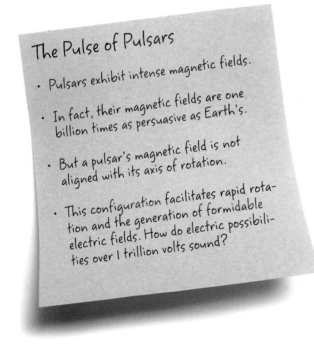

The Pulse of Pulsars

- Pulsars exhibit intense magnetic fields.

- In fact, their magnetic fields are one billion times as persuasive as Earth's.

- But a pulsar's magnetic field is not aligned with its axis of rotation.

- This configuration facilitates rapid rotation and the generation of formidable electric fields. How do electric possibilities over 1 trillion volts sound?

Crab Pulsar

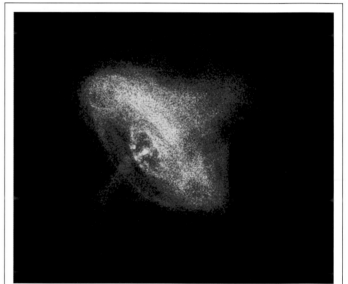

- The Crab nebula in Taurus showcases one of the most prominent examples of a pulsar. It is appropriately named the "Crab pulsar."

- It is the middle star in the Crab nebula.

- This neutron star—pulsar—is believed to be relatively

young in stellar terms, having been born in the wake of Supernova 1054, observed on Earth in 1054.

- The X-rays of the Crab pulsar are visible when the spinning neutron star's hot spots are in line with our view here on Earth.

112

a pulsar would hit it with the brute force of several hundred hydrogen bombs. With these rather startling characteristics on display, it's little wonder that pulsars furnish scientists with invaluable lessons on the mercurial behavior of gravity.

When a beam of light radiating from a pulsar sweeps over Earth during its high-speed revolutions, it appears from observers' perspectives to pulsate—on, off, and then on again. Astronomers have likened a pulsar's actions to a lighthouse—that is, intermittent light flashes as opposed to uninterrupted rays of light. Pulsars are considered by scientists to be one of the most mysterious celestial bodies in the universe. They are also among the brightest objects in the vast regions of outer space. However, it isn't just a pulsar's intensity that fascinates observers of the night sky.

Recently NASA, employing its Fermi-Gamma-ray space telescope, identified several pulsars that spin even faster than average pulsars and emit high-energy gamma rays into space. These "millisecond" pulsars, as they are popularly known, spin with the ferocity of our kitchen blenders on high speed.

Pulsar Nebula PSR B0540-69

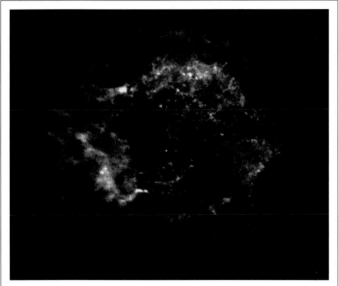

- PSR B0540-69 is considered a virtual twin of the Crab pulsar in the Crab nebula.

- The duo's energy and wind behaviors are quite comparable.

- PSR B0540-69 is considered young in stellar time—just behind the Crab pulsar in reputed age. Among the isolated pulsars currently known, it is a mere child.

- Like the Crab pulsar, PSR B0540-69 calls home the Large Magellanic Cloud.

Vela Pulsar

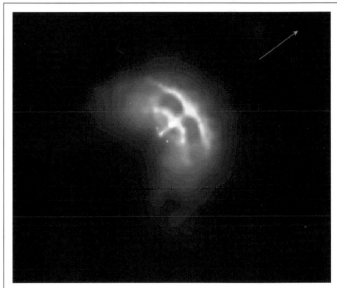

- The Vela pulsar is sometimes referred to as the "Vega pulsar." It is located in the constellation Vela.

- Its pulsating and glowing orange image reveals the effects of the pulsar wind.

- This gamma-emitting space body is likely a deriva-tive of the Vela supernova remnant: remains from a supernova event that occurred approximately twelve thousand years ago.

- The Vela pulsar is barely 12 miles (19.3 kilometers) in diameter, but it rotates at approximately ten times a second.

MAGNETARS
Magnetars' magnetic fields are a thousand trillion times as strong as Earth's magnetic field

Like pulsars, magnetars are neutron stars. Magnetars are in essence offshoots of pulsars. They kick everything up a notch, so to speak, including their spinning ways and formidable magnetic capabilities.

From scientists' perspectives, magnetars' unusual conduct supplies them with yet another pathway for stars to emit light and shine brightly in the far reaches of outer space. That is, magnetars do not cast light into the interstellar medium because of nuclear fusion, which is the familiar source of countless stars' radiance. Nor do they achieve their ultimate energy through a superfast spinning rotation, the traditional pulsar's bread and butter. Rather, magnetars' magnetic fields,

The Magnetar Magnet

- Relative to what we know on Earth, pulsars exhibit superstrong magnetic properties.

- But one billion times as potent as our planet's magnetic capacity is not one thousand trillion times, which is what magnetars are believed to enjoy.

- One thousand trillion, by the way, is a numeral one followed by fifteen zeroes.

- Magnetars are indisputably the universe's most extreme magnets.

Birth of a Magnetar

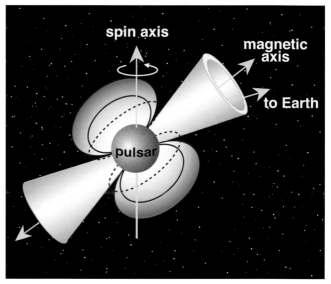

- A just-formed neutron core of the pulsar spins counter-clockwise on its axis. But, plainly, its magnetic axis is not in concert with its rotation.

- Extreme magnetic fields are the byproduct of rapid spinning on its axis.

- Scientists are nonetheless baffled by the super-strength of the magnetic fields found in magnetars.

- It is impossible to duplicate such magnetic capacities in laboratories. This is why magnetars continue to amaze all who study them.

which are exceedingly potent, are the underpinning of their luminosity.

Estimated at 1 million billion Gauss, magnetars' magnetic muscle is quite phenomenal. (Gauss, often abbreviated as G, is a unit of magnetic strength.) For comparative purposes, the sun's magnetic field is merely 5 Gauss. It is felt that a magnetar's ultramagnetic field wreaks havoc of sorts on its outer crust, which is an otherwise durable feature on the multiple millions of neutron stars observed by astronomers. That is, a magnetar's magnetic force warps and eventually breaks apart the crust, spawning ferocious seismic waves that energize space matter beyond the surface layers of this unusual star. In other words, the light generated by a magnetar comes to pass in an especially circuitous way.

Although scientists assume there are likely many more of them out there in the vast expanse of the universe, they have officially identified only ten magnetars to date in our parent galaxy, the Milky Way, and all of the other visually explored galaxies as well.

Magnetar SGR 1900+14

- The magnetar SGR 1900+14, emitting gamma rays, can be found in the constellation Aquila.

- It is believed to be a remnant from a relatively recent—1,500 years ago or so—supernova event.

- NASA's Spitzer space telescope has recently detected a glowing ring surrounding this magnetar.

- Astronomers estimate SGR 1900+14's magnetic field to be 500 trillion times stronger than the comparable field here on Earth.

Magnetar Magic

- The illustration of the magnetar reveals its core, the remains of a past star. But not just any star: a star with tremendous mass that met its stellar end in grand fashion.

- The magnetar's extraordinarily rapid rotation is generating magnetic field lines that appear to be encircling its nucleus.

- The surface of the magnetar is incredibly hot.

- The magnetar spins on its axis and flashes both hard X-rays and soft gamma rays.

BLACK HOLES

With incredible gravitational muscle, peculiar black holes are the remnants of supermassive stars

Black holes are puzzling areas in outer space that continually intrigue both professional and amateur astronomers alike. For a long time, black holes were considered areas of inter-stellar mayhem and destruction. That is, clandestine cosmic monsters gobbling up all forms of matter into their shadowy and insatiable vortexes. However, with increasing amounts

of scientific information barreling down the pike, black holes are now seen in a different light, if you will. Today they are widely felt to be important cornerstones in both the origins and development of the universe.

Essentially, black holes in space are expanses where the tug of gravity is so extreme that nothing—not even light—can

Elliptical Galaxy M87

- The giant elliptical galaxy M87 is believed to accommodate a black hole of significant proportions.

- In fact, scientists surmise that a black hole of 6.4 billion solar masses is likely lurking in this galaxy's epicenter.

- The diagonal streak emanating from the galaxy's nucleus, approximately 6,500 light-years across, quite possibly originates from this black hole.

Elliptical Galaxy NGC 1316

- NGC 1316 is a sprawling elliptical galaxy. It is sometimes referred to as a "lenticular galaxy": a hybrid of both spiral and elliptical galaxies.

- NGC 1316 is located in the constellation Fornax in what is known as the "Fornax cluster of galaxies."

- Scientist Francois Schweizer extensively surveyed this galaxy in the late 1970s and concluded that it is the off-shoot of multiple galaxies that merged through time.

- He also suspected a super-massive black hole at the galaxy's center.

ever escape. But despite their at-once indiscernible and volatile temperaments, scientists know that black holes exist throughout the universe because they accrete matter and leave the telltale signs of having done so. Kinetic energy generates heat that ionizes atoms, which in turn release X-rays that can be observed by scientists.

More specifically, black holes are the potential endgames for enormous stars breathing their last. Stars with masses more than a million times that of the sun, in the aftermath of supernova explosions, sometimes become black hole remnants.

Courtesy of its relatively small size, the sun could never form a black hole. But when a much bigger dying star than our sun sheds its mass, sans any counteracting gravitational forces, it implodes to a level of zero volume but incalculable density. This progression engenders a region of space called "singularity," where the forces of gravity completely overwhelm all comers.

Black holes are believed to reside in the center of galaxies throughout the universe. Although they can come in all sizes, massive black holes are thought to be the gravitational behemoths at the heart of galaxies, including the Milky Way.

Sombrero Galaxy and Black Hole

- The Sombrero galaxy, M104, is a celestial sight for sore eyes located in the zodiac constellation Virgo.

- It is an unbarred spiral galaxy approximately 29 million light-years away from Earth.

- Scientists have detected what they feel is a consider-able black hole in the center of the Sombrero galaxy.

- Calculating the speeds of the stars orbiting its center, it is surmised that this black hole contains the mass of one billion suns. This would make the Sombrero galaxy's black hole one of the most enormous.

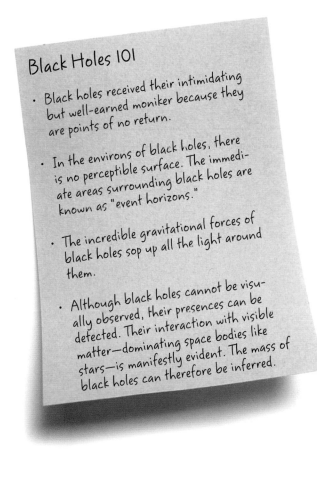

Black Holes 101

- Black holes received their intimidating but well-earned moniker because they are points of no return.

- In the environs of black holes, there is no perceptible surface. The immediate areas surrounding black holes are known as "event horizons."

- The incredible gravitational forces of black holes sop up all the light around them.

- Although black holes cannot be visually observed, their presences can be detected. Their interaction with visible matter—dominating space bodies like stars—is manifestly evident. The mass of black holes can therefore be inferred.

NOVAE & SUPERNOVAE

Nova and supernova events are colorful interstellar explosions, but they are very different

Both nova and supernova celestial occurrences involve interstellar explosions. And because of their designations, the two outer-space events are frequently confused as being one and the same or closely related. However, they are vastly different animals.

Novae (plural) take place when white dwarf stars, which are

in effect dead stars that have shed their outer layers, endure as simply extraordinarily hot inert cores and interact with companion stars. Stars frequently pair up and orbit one another during their lifetimes. They are known as "binary stars." But sometimes a white dwarf grabs hold of hydrogen being set free from its red giant companion. "Red giant" is the star classification

Nova Cygni 1992

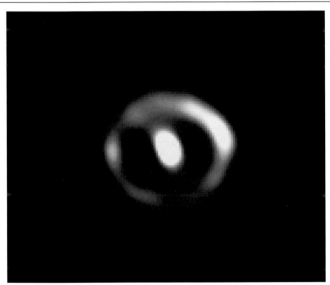

- In 1992 a spectacular stellar explosion occurred in the constellation Cygnus: a nova event.

- Nova Cygni 1992 was the brightest nova in recent memory. At its peak, it could be seen without a telescope.

- The Hubble space telescope captured images of the white dwarf star's expanding outer layers.

- The ring around it is evidence of the superhot gas blasted into the ether by the explosion.

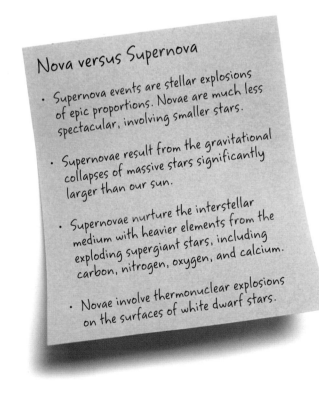

Nova versus Supernova

- Supernova events are stellar explosions of epic proportions. Novae are much less spectacular, involving smaller stars.

- Supernovae result from the gravitational collapses of massive stars significantly larger than our sun.

- Supernovae nurture the interstellar medium with heavier elements from the exploding supergiant stars, including carbon, nitrogen, oxygen, and calcium.

- Novae involve thermonuclear explosions on the surfaces of white dwarf stars.

immediately preceding white dwarf rigor mortis. That is, when the star is releasing both its outer layers and remaining surface hydrogen along with it. In other words, when hydrogen pours off the red giant and reaches the white dwarf's smoldering core—boom! This thermonuclear explosion supplies an instantaneous and incredible increase in the white dwarf's luminosity, which was theretofore faint. The word *nova* is Latin for "new star."

Conversely, supernova explosions involve different stellar conditions altogether and supply vastly dissimilar outcomes as well. When very old and very big, stars reach their natural ends—that is, they cease to generate nuclear fusion because of depleted hydrogen—they collapse into themselves and prompt supernova explosions. On occasion, a supernova event gleams with the intensity of more than ten billion of our suns. Such radioactive fireworks have actually generated ample lighting to obscure entire galaxies.

Supernovae are considered among the most spectacular outer-space happenings, displaying both kaleidoscopes of colors and extraordinary illumination. But supernova space events are not only impressive visuals.

Supernova Kepler

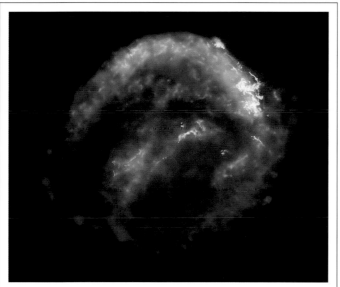

- The last verifiable supernova event in the Milky Way occurred in 1604. SN 1604, or Supernova Kepler, was observed in the constellation Ophiuchus.

- Remnants of Supernova Kepler are presently visible as a bright veil of gas and dust spanning 14 light-years in diameter.

- Johannes Kepler chanced to see this event with the naked eye.

- At the peak of this stellar explosion, SN 1604 was the brightest object in the night sky.

Type II Supernova

- The most spectacular supernova events are called "Type II." These stellar moments occur when massive stars exhaust their fuel sources.

- The stars implode when they can no longer be sustained by nuclear fusion.

- Their cores, massive from a buildup of iron, collapse in a spectacular explosion.

- Some well-known stars have been deemed candidates for the supernova events of tomorrow—hundreds of thousands to millions of years from now—including Betelgeuse, Antares, and Spica.

THE SUN

Our sun is an average-sized star that is responsible for all life on Earth

The sun is the most famous star in the universe. It has been known for as long as recorded time, so history records no discoverer of the star that is wholly responsible for life on Earth.

Like countless billions of other stars, the sun is a shimmering orb consisting of myriad gases that is generating energy via nuclear reactions in its searing and lively center. Although

the sun is rather considerable in size, it is nonetheless considered rather average-sized in the pantheon of stars.

The sun's visible layer—what we see when we look up to in the sky—is called its "photosphere." The temperature there is 11,000°F (6,000°C). The photosphere is often referred to as the sun's "surface," but it is not truly a surface. That is, the sun

A New Day

- The sunrise is a sighting to behold. Most people never see one because they are asleep at the time.

- The official definition of "sunrise"—the exact time varies slightly from day to day as the calendar advances—is when the sun's leading edge materializes immediately above the horizon.

- "Dawn" is not synonymous with "sunrise."

- Dawn occurs when the early morning sky first brightens, before the sun is sighted on the horizon.

Sunrays Get Sidetracked

- When the sun's rays flow through openings in the clouds, or between countless earthly objects, they create warm and reassuring visuals.

- Sunrays that appear to streak from a specific location in the sky are called "crepuscular rays."

- Sunrays that jet out from low fair-weather clouds are sometimes called "Jacob's Ladder."

- The sun's rays are frequently scattered by Earth's layers of atmosphere and particles in the air.

consists of gases, not solid materials to establish what you could call a bona fide surface. You couldn't actually walk on the sun, but then you wouldn't want to anyway.

The sun's core is where the real action and the hottest temperatures are: 27 million degrees Fahrenheit (15 million degrees Centigrade). It is estimated that, via nuclear fusion, the sun converts more than 700 million tons of hydrogen into helium every second of every day. This ongoing process is what generates energy and indeed the heat that eventually wends its way down to Earth.

ZOOM

Our sun is fortunately what is known as a "second-generation star." That is, it is not as old as the presumed age of the universe but merely 4.6 billion years old, with a heaping helping of living still left to do. Scientists surmise that the sun has sufficient hydrogen to fuel its existence for billions of years before it becomes a red giant star and then a white dwarf star.

Bright Sunshine, Blue Skies

- The combination of bright sunshine and blue skies is a most welcome weather one-two punch.

- Assuming that it is undisturbed, light hurries through space in a veritable straight line. It is blown off course only when it encounters obstacles.

- Earth's atmosphere is one such obstacle that interferes with light sources.

- Blue light is absorbed by molecules of gas in the atmosphere, strewn in all directions, and the architects of our blue skies.

Another Day in the Books

- Sunsets are ordinarily more spectacular than sunrises.

- This isn't because the sun's composition and color scheme are any different at the end of the day.

- Its rich orange and red hues are the byproduct of dust particles in the atmosphere, including pollutants, which distort actual colors of light.

- There are usually more particles in the air during the evening than in the morning, impacting the visual color of the sun from our earthly perspective.

SOLAR SYSTEM CHOREOGRAPHER

In a sprawling galaxy, our solar system is a tiny niche dominated by the sun

The sun is at the heart of what we call a "solar system." That is, a region in space dominated by a star with myriad objects of less mass orbiting around it. This dominion is directly due to the star's persuasive gravitational sway. It is believed that both our sun and solar system were formed roughly 4.6 billion years ago, the consequence of an enormous molecular cloud crumbling to pieces in rather grand fashion. Although not considered a massive star in the sprawling stellar family of stars, the sun nonetheless commands the mass in our solar system, accounting for 99 percent of its total. This absolute hegemony is what draws planets and other space bodies into its orbit.

Solar System Big Wheel

- Courtesy of the sun's mass—almost 99 percent of our entire solar system—its gravitational influence dominates all objects around it.

- Its commanding presence reaches beyond Neptune's orbit to a place in space known as the "heliopause."

- At this point in our solar system, the sun's solar wind begins to wane.

- Outside of what is called the "heliosphere," the sun's force is not an issue anymore, but another star's gravitational magnetism is. Where this area begins is where our solar system ends.

Big Sun, Small Sun

- From our vantage point, the sun is huge. However, from far away—outside of our solar system, for instance— it would appear like just another star in the night sky. One among the countless stars that blankets the celestial sphere.

- In the fraternity of stars, our sun is a second-generation member.

- That is, the sun—at 4.5 billion to 5 billion years old— is not as old as the universe.

- First-generation stars are believed to be as old as the universe itself.

In fact, the eight planets—Mercury, Venus, Earth, Mars, Jupiter, Saturn, Neptune, and Uranus—orbit in concert with the sun's rotation, which is counterclockwise if observed from its north pole. A similar orbiting pattern holds true for most other space objects in our solar system, including dwarf planets such as Pluto and Ceres.

Interestingly, the notion of a solar system wasn't publicly bandied about in scientific circles until 1543, when Nicolaus Copernicus published his heliocentric theory. Copernicus placed the sun at the center of his model of the celestial beyond, with all else revolving around it in a neatly ordered system. Seventeenth-century astronomers Galileo, Kepler, and Newton contributed mightily to the acceptance of the heliocentric model. Their work in the fledgling science of physics made heliocentric opinion both logical and increasingly acceptable to the masses. Finally, our solar system was out of the celestial closet.

From where we sit here on Earth, the sun rises in the east and sets in the west. Its regular course, known as the "ecliptic," explains the seasons of the year.

Harmonious Stellar Pair: Sun and Earth

- The sun's gravity ensures that Earth orbits around it in a soothingly conventional course.

- Earth and the sun are believed to be close in age.

- The oldest rocks found on our home planet are four billion years old, and the sun's presumed age is just several million years older than that.

- Ancient Greeks named the sun "Helios," whereas the Romans named it "Sol."

Planets Orbiting the Sun

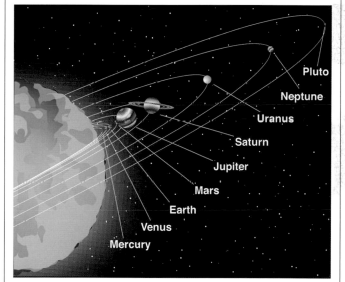

- The sun dominates the region where its gravitational sway and solar wind are felt. This is our solar system with eight officially recognized planets.

- The planet Mercury orbits nearest to the sun, whereas the planet Neptune is the farthest away.

- Pluto, which was once considered a planet—and is now considered a "dwarf planet"—orbits a greater distance away than Neptune.

- Most other objects in our solar system—including comets, asteroids, and meteoroids—have little choice but to orbit the sun.

REGIONS OF THE SUN
The sun has a hot core, gaseous surface, and millions of miles of atmosphere

The sun has three major regions outside its hot core: the photosphere, chromosphere, and corona. The photosphere is both its deepest area and what is generally visible to us when we observe the sun in the sky. For lack of a more apt depiction, the photosphere is the sun's surface. However, although the sun is made up of similar elements as Earth, its extreme temperatures do not permit them to exist in solid states.

The photosphere is, therefore, completely gaseous. Scientific observers can penetrate no farther than this surface layer of the sun, which grows increasingly dense as you move toward the solar core.

The region of the sun immediately above the photosphere

The Sun Inside and Out

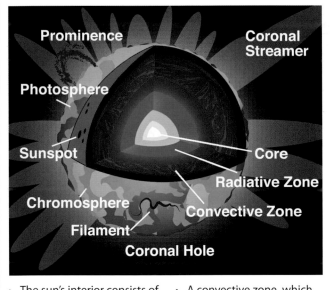

- The sun's interior consists of gas from its core to surface. It is so hot that solids do not exist. Whereas Earth's innards contain molten rock, the sun has no such thing.

- Energy from the superhot sun's core wends its way through a radiative zone.

- A convective zone, which marks the outer 15 percent of the sun's radius, facilitates the transfer of its heat and energy to the surface and beyond.

- The gaseous photosphere, for lack of a more apt description, is the surface of the sun.

The Photosphere

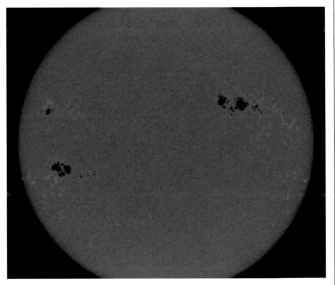

- The sun's photosphere is what we observe when checking it out in the sky.

- The photosphere is not a surface layer per se. There is no solid core surrounding the sun—only gases. However, the gases that constitute the photosphere are very dense. We cannot see past the photosphere to the sun's insides.

- The photosphere is approximately 300 miles (500 kilometers) thick.

- The sun's photosphere temperature is believed to be as high as 5,840 degrees Kelvin.

is its first layer of atmosphere called the "chromosphere." Paradoxically, the temperature rises in this immediate layer above the sun's *surface*. This is contrary to the falling temperatures in the remainder of the sun as you venture away from its scorching core.

The chromosphere ultimately fuses with the sun's corona, which is its remotest boundary. The corona literally stretches millions of miles beyond the photosphere and into the interstellar medium. The corona's vividness is typically obscured by the sun's photosphere, which is even brighter.

Granulation Pattern in the Photosphere

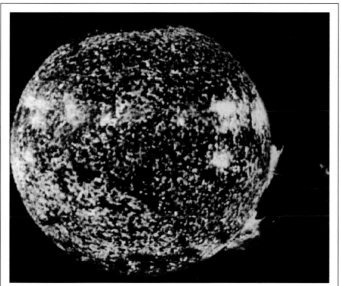

- Granulation patterns are regularly spotted on the photosphere.

- Hotter gases encountering relatively cooler regions of the photosphere cause the patterns. A temperature difference of 1,000 degrees Kelvin to 1,500 degrees Kelvin cooler is enough to inspire these changes.

- The lighter color in the granulation pattern is evidence of cooler gases rising from the convection zone.

- The darker lines are known as "intergranular lanes."

The Chromosphere

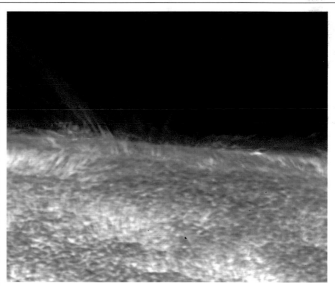

- The area of the sun located above the photosphere is called the "chromosphere."

- It is estimated to span 1,200 to 1,800 miles (2,000 to 3,000 kilometers) off the surface.

- The sun's chromosphere is not visible during the daylight hours.

- The sun's chromosphere and radiant white corona come into view only during total solar eclipses.

SUNSPOTS

Sunspots are tiny black spots on the otherwise reliably white surface of the sun

Sunspots are small dark splotches on the sun's surface. However, they are considered small only relative to the size of the sun, which spans 865,000 miles (1,393,000 kilometers) in diameter. Sunspots, for instance, have been measured on average at 2,500 to 10,000 miles (4,000 to 16,000 kilometers), with some sprawling outward as much as 50,000 miles (80,000 kilometers).

These darkened blemishes on the sun's face signify cooler areas than the overall photosphere temperatures, which run in the neighborhood of 11,000°F (6,000°C). However, we are talking about only several thousand or so degrees cooler. So, a sunspot may register a temperature at 6,000°F (3,300°C) or thereabouts, which is still pretty hot.

Colossal Sunspot

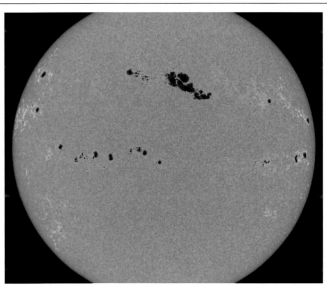

- Huge and widespread sunspots were spotted, if you will, on the sun's surface in 2001.

- Sunspots are areas on the sun's photosphere with very strong magnetic fields,that showcase cooler temperatures than those of the overall surface.

- The temperature differences could more aptly be described as "relatively cooler." That is, the cooler regions are still flaunting thermometer readings of 3,500° to 4,500°K.

- As is always the case with sunspots, these observations were temporary.

Sunspots at Solar Minimum

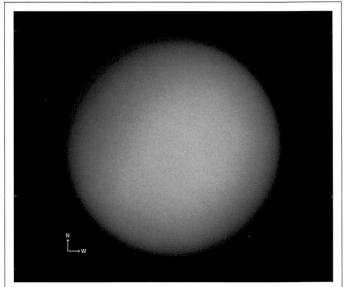

- A more inactive sunspot period is known as a "solar minimum." During this stretch of time, sunspots are much less frequent.

- Although the cycles average approximately eleven years, there have been decidedly shorter and longer cycles.

- A lengthy period in history, which has been coined the "Little Ice Age," spanned the years 1645–1715. During this time weather on Earth was much colder than normal.

- Some scientists feel that this weather anomaly was directly related to solar behavior.

Extreme magnetic activity, which is both compelling and commonplace on the sun's surface, triggers these recurring sunspots. Perpetual magnetic machinations sometimes impede convection. That is, the stream of replenishing hot gases from the sun's center is altered in some fashion. The sun's interior temperatures have been estimated at 29 million degrees Fahrenheit (16 million degrees Centigrade). Indeed, this obstructing event breeds little pockets of lower temperatures that appear black in color against the otherwise-uniformly white backdrop of the photosphere.

Individual sunspots are not, however, permanent features of the photosphere. Some sunspots have been known to appear and then disappear in a matter of days, whereas most last a month or so, which is a typical solar rotation.

Scientists who study sunspots refer to their interior sections as the "umbra." This is at once the darkest and relatively coolest area of a sunspot. The exterior is called the "penumbra," which is visibly less dark than the umbra. Typically sunspots crop up in pairs.

Sunspots at Solar Maximum

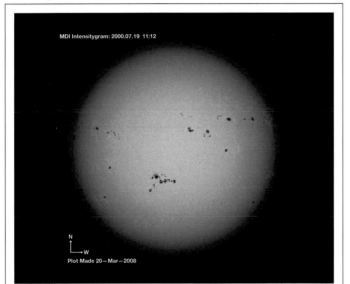

Sunspot Close-Up

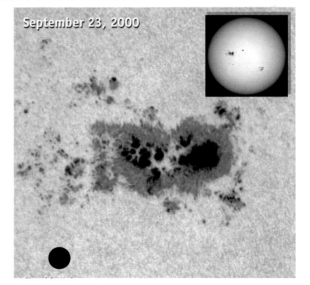

- Sunspots occur frequently but run in high and low cycles.

- An active sunspot period is known as "solar maximum." During this span of time, sunspots are more likely and more numerous.

- The average length of these cycles runs eleven years or so—maximum and then minimum—but they are not set in stone.

- The intensity of each solar maximum cycle varies greatly. Some cycles are more prolific and inspire more solar flares than do other cycles.

- Sunspots resemble acne on the face of the sun. A bad case occurs during solar maximum.

- Sunspots coincide with solar flares.

- Solar flares are extreme eruptions of ionized materials that break away from the sun's gravity and shoot through space with alacrity.

- Upon reaching Earth, these flares sometimes wreak havoc on radio communications and electrical and telephone lines. They also are behind the colorful aurorae—the Northern and Southern Lights.

SOLAR WIND

Perpetually flowing off the sun, fast-blowing solar wind continues to shape our solar system

Every now and then we hear about glitches in telecommunications that are the direct consequence of the sun's unremitting solar wind. Solar wind has inspired geomagnetic storms in the Earth's atmosphere, which have been known to wreak havoc with our power grids. Indeed, solar wind makes its presence felt in a variety of locales and ways.

Solar wind is chiefly electrons and protons flowing off the sun's corona. That is, 1 million tons per second of material are streaming from the sun's outermost atmospheric layer into interplanetary space with abandon. Solar wind travels at an average speed of 250 miles per second (400 kilometers) and has been measured as high as 500 miles per second

Solar Wind Blowing Off the Sun

- The solar wind streams off the sun's corona—its outermost atmosphere—into the ether of space.

- The solar wind propels plasma—highly energized electrons and protons—throughout the entire solar system.

- Solar wind blows as fast as 500 miles (800 kilometers) per second.

- And the temperatures of the particles perpetually coursing off the sun are in the neighborhood of 1,800,000,000°F (1,000,000,000°C).

Comet Makeover

- When in the immediate environs of the sun, Comet Hale-Bopp is shaped by the solar wind.

- By melting the comet's frozen core, the superhot and fearsome solar wind fashions its tail of gas and dust.

- The solar wind ensures

that Hale-Bopp's tail trails away from both its core and indeed the sun.

- All comets with perceptible tails owe the sun's radiation and solar wind a debt of gratitude because after they leave the environs of the sun, their tails gradually dissipate.

(800 kilometers). Because it is so indefatigable, its impact is widespread and far-reaching. Celestial objects have been shaped and reshaped through time because of intense solar wind battering. Earth has a protective atmosphere that prevents the worst of solar wind mauling, although our planet is not immune from its effects. Interestingly, the solar wind's plasma is quite thin in the environs of Earth, and although it's blowing with genuine ferocity, it wouldn't even mess up the hair on your head on a solar windy day.

The solar wind has created what is known as a "heliosphere,"

a protective bubble that accommodates our solar system. This bubble is the byproduct of the highly charged particles of the solar wind meeting and greeting the wandering and teeming gases that make up the interstellar medium. Solar wind is also responsible for what we on Earth deem "space weather" and the spectacular light shows known as "aurorae," or more colloquially, the Northern and Southern Lights. We also owe a debt of gratitude to the solar wind for supplying comets with their vivid plasma tails.

Moon Walking

- Astronauts walking on the barren surface of the moon witnessed firsthand the consequences of solar wind.

- With no atmosphere or magnetic field to protect it from the relentless and harsh interplanetary solar wind, the moon is an austere environment.

- The solar wind's ions are ingrained in the moon's topsoil, which is called its "regolith."

- Samples of the moon's regolith, brought back to Earth by *Apollo* missions, have proved invaluable in the study of solar wind.

Solar Wind Composition Experiment

- Multiple *Apollo* missions—*11*, *12*, *14*, *15*, and *16*—employed what was dubbed the "solar wind composition experiment."

- Solar wind particles rain down on the moon's surface all the time.

- These missions erected

sheets of foil, which were exposed to the solar wind for varying periods of time. The experiments gathered solar wind particles that lodged in the foil.

- Earth's magnetic field thwarts the solar wind's charged particles from reaching the surface.

COLOR OF THE SUN
The sun appears to us in diverse hues, but it is in truth one color

When we glance up into the sky during a bright sunny day, the sun appears yellowish in color. Sometimes it's definitively yellow, whereas on other occasions it's seemingly a blend of yellow and white. And then there are sunrises and sunsets, when the sun exudes an intense red hue or an intoxicating blend of orange and red.

These variations in the sun's appearance, as seen here on Earth, occur not because it is varying its color scheme from morning through evening and from day to day. No, the sun is, in fact, one solid, unchanging color: white. This is the color you would see from a catbird's seat in the recesses of outer space, where there is no atmosphere filtering and dispersing its uniform color.

Earth's layers of atmosphere scatter sunlight before it reaches the surface and our eyes. More specifically, the atmosphere diminishes shorter wavelength light—that is,

True Colors

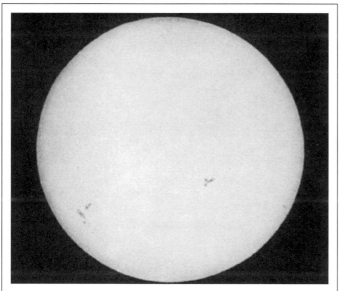

- The color of an individual star is based on its temperature. Stars aren't uniform colors. Our sun has a color, but that color is not what we see.

- Our sun is actually white. Our planet's atmospheric layers reorganize the sun's colors. This is why we see the sun in various hues.

- Sometimes it's yellow in the skies. At sunset, it's often a fiery red or orange.

- Stars like our sun are white based on their temperatures of approximately 6,000°K.

Whitish-Colored Sun in the High Sky

- Sometimes the sun appears whiter than any other color during the daytime. This appearance is nearer its genuine color.

- When the day comes when our sun cools down in temperature—below 3,500°K—it will officially be a red giant star and be red, not white, in color.

- The hottest stars, above 10,000°K, are blue.

- The temperatures of stars encourage colored photons, which maintain varying amounts of energy, to reveal themselves: white, red, blue, and so forth.

the blues and violets. And with these colors eliminated from the light spectrum, we see the sun as yellow. And the same atmosphere conditions are at work during sunrises and sunsets, reflecting and scattering light. The difference in colors at both these early and late hours can be attributed to the sun's positioning near the horizon. That is, the sun's light passes through more atmosphere at these moments, which occasion additional distortions of the literal color of the sun.

.... GREEN ● LIGHT ....

If you haven't witnessed a sunrise or sunset, you don't know what you are missing. Although the sun's actual color is pure white, sunrises and sunsets immensely benefit from the layers of our atmosphere, which distort the light in our favor. Also, when the sun is on the horizon with its rays passing through additional atmosphere, it is safe to observe, which is in stark contrast to when it is high in the sky.

Yellow-Hued Sun

- The sun in the high sky typically appears yellow or yellowish in color.

- This common view is courtesy of Earth's atmosphere, which scatters sunlight.

- Our atmosphere eliminates what is considered shorter wavelength light—blue and violet—from the color spectrum. This makes the sun appear yellow or yellowish.

- What we see are heavily filtered colors, which from our perspective are a good thing. The sun's hues are given both more variation and oomph.

Reddish Skies at Sunset

- The setting sun is the most widely sought out and photographed.

- At sunset, with the sun close to the horizon, its color is often flaming red or orange-red. It appears as a ball of fire in the sky.

- During sunset, the sun is located on the horizon. Its lighting must travel through more atmosphere than when it's high in the sky.

- More layers of atmosphere to plow through mean more scattering of light and a more spectacular color scheme than the sun's pure white.

131

WHAT ARE CONSTELLATIONS?

Constellations are unique patterns of stars assigned specific areas in space and named

Many of the constellations of stars on the celestial sphere are well known. From Canis Major the Great Dog—housing the brightest star in the night sky, Sirius—to Sagittarius the Archer, one of the zodiac constellations, these star-laden regions of space get a fair amount of attention.

Constellations are merely named patterns of stars in the night sky. By dissecting the vast expanses of outer space, constellations assume a role akin to individual state boundaries in America or country boundaries on continents. They enable night-sky observers to locate stars and other celestial objects by essentially visiting space neighborhoods.

Constellations supply a measure of order to the night sky.

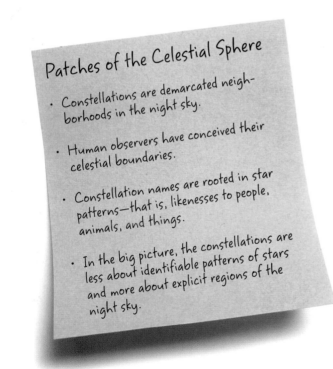

Patches of the Celestial Sphere

- Constellations are demarcated neighborhoods in the night sky.

- Human observers have conceived their celestial boundaries.

- Constellation names are rooted in star patterns—that is, likenesses to people, animals, and things.

- In the big picture, the constellations are less about identifiable patterns of stars and more about explicit regions of the night sky.

Hi, Neighbors

- Canis Major the Great Dog accommodates the brightest star in the night sky, Sirius.

- Sirius, nicknamed the "Dog Star," is also part of the asterism known as the "Winter Triangle."

- Canis Minor, appreciably north of Canis Major, is known as the "Lesser Dog."

- In Greek mythology, the imagined two dogs in the night sky are nipping at the heels of Orion the Hunter.

They also facilitate the naming of celestial bodies, which are often furnished monikers based on their constellation of residence. It's like New York City receiving its name for its state of residence.

The origins of many constellation designations date back to ancient times. Starwatchers from yesteryear observed bright stars that appeared close together, even though they were actually long distances from one another. Nonetheless, these starwatchers divided the night sky into various named regions based on these star groupings.

Big Dipper

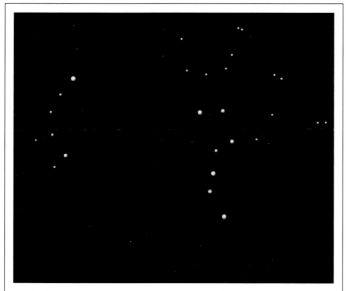

- The Big Dipper is the most familiar asterism in the night sky and often used as a star-hopping point of reference in the Northern Hemisphere.

- An asterism is not a constellation in and of itself but rather a pattern of stars within a constellation or in multiple constellations.

- Asterisms assist stargazers in navigating the sprawling night sky.

- Other well-known and sought-out asterisms include the Great Square of Pegasus, Orion's Belt, the Summer Triangle, and the Teapot in Sagittarius.

Connect the Stars

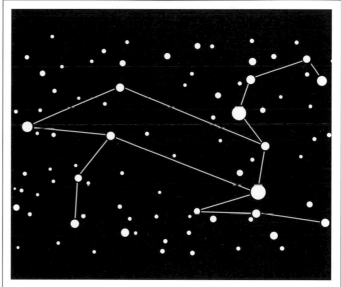

- Leo the Lion was one of the earliest known constellations. That is, a specific area of the sky identified by ancient civilizations.

- The stars in this, the fifth zodiac constellation, highlight what may, or may not, resemble a crouching lion gazing westward.

- As with many of the constellations, their alleged contours, or lack thereof, are in the eyes of the stargazer.

- Leo is prominent in the Northern Hemisphere's spring skies and Southern Hemisphere's autumn skies.

133

EIGHTY-EIGHT CONSTELLATIONS

The officially recognized constellations include classic and modern

The International Astronomical Union (IAU) officially recognizes a total of eighty-eight constellations of stars. Forty-eight constellations are considered the "classics." That is, constellation borders that were first recognized by ancient Greek and Roman civilizations. However, after Nicolaus Copernicus's pioneering work in the field, post-Copernican astronomers of the 1700s and 1800s identified areas in space that were not addressed by the ancient cultures, notably areas of the Southern Hemisphere skies that were never visible in the Northern Hemisphere. Remember that the ancient Greeks and Romans surveyed the night skies from only their lands in the Northern Hemisphere. There were, in fact, portions of the celestial sphere that merited both a look-see and constellation stamps of their own.

In addition to areas of the night sky previously unexplored, astronomers in the eighteenth and nineteenth centuries

Old and New

- The Greek mathematician and astronomer Claudius Ptolemaeus, better known as "Ptolemy," recorded the names and locations of forty-eight constellations, which are now referred to as the "classics."

- One of the first in-depth astronomical dissertations was Ptolemy's work, *Almagest* published in the second century A.D.

- He explained both where the constellations are and what could be found in them.

- The forty subsequently recognized constellations are known as the "modern constellations."

Space Neighborhoods on the Celestial Sphere

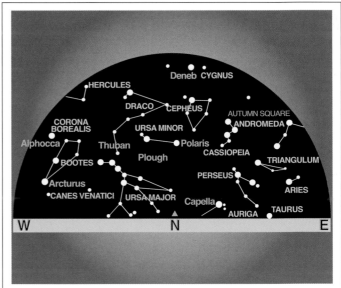

- At any given moment, the nighttime celestial sphere furnishes you with a snapshot of numerous constellations of stars.

- No matter where you are or the time of year, the night sky is teeming with stars hanging out in their celestial neighborhoods.

- This picture changes as each night wears on, with constellations seemingly rising in the east and setting in the west.

- More fundamentally, this same picture window remakes itself with the changing seasons.

found patterns of stars in between already designated constellations that they felt merited homes of their own. That is, orphan stars that belonged in newly assigned constellations. And so new constellations were born. The postclassic constellations are known as "modern" constellations. There are forty modern constellations, which have been added to the original forty-eight, enabling you to peruse the entire night sky by neighborhood, if you will, with no regions excluded.

Most of the constellations of stars appear to rise in the east and set in the west—diurnal motion. There are also constellations that do not ever "rise" or "set." These are the circumpolar constellations—Ursa Major, Ursa Minor, Draco, Cassiopeia, and Cepheus—that are visible throughout the year. Instead of gliding across the sky from east to west, circumpolar constellations revolve around a fixed point in space—the celestial north pole. And depending on their prime visibility in the night sky, constellations one and all are likewise categorized seasonally.

Constellation Etymology

Lupus

- Many of the eighty-eight constellations—notably the classic forty-eight—have interesting and even Byzantine etymologies.

- Lupus the Wolf, an obscure constellation in the southern sky, was not immediately identified as a wolf. It wasn't until the sixteenth century

that it got its moniker.

- Prior to that time, Lupus was considered a generic wild animal.

- The neighboring Centaurus the Centaur is presumed to be carrying this slain animal in his arms.

Camelopardalis the Giraffe

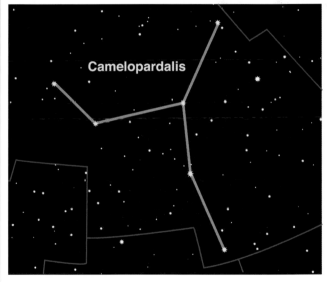

Camelopardalis

- Camelopardalis the Giraffe is a sprawling but largely dim constellation in the northern sky.

- In order to make out the constellation's giraffe outline, you will have to locate many faint stars within its borders. And even then, it might be straining credulity

to envision a connect-the-dots giraffe among the stars.

- Officially recognized in the seventeenth century, Camelopardalis is among the modern constellations. The ancient Greeks viewed this sliver of the night sky as barren.

WINTER CONSTELLATIONS

Prominent winter constellations include Canis Major, Eridanus, Gemini, Orion, and Taurus

Constellations of stars are usually classified as seasonal. That is, there are times during the year when the individual constellations are visible for optimum observation. In all instances, it matters where you are. Even so-called winter constellations are not evident to everybody everywhere during wintertime.

The winter constellations of the Northern Hemisphere include Auriga, Canis Major, Canis Minor, Cetus, Crater, Eridanus, Gemini, Lynx, Orion, Perseus, Sextans, and Taurus.

Winter skies are the clearest of any of the seasons. The generally colder and drier air plays fewer hazing pranks on the atmosphere. If you can withstand the colder temperatures, it's recommended that you wade into the winter night skies.

Canis Major the Big Dog

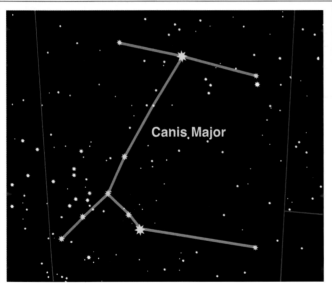

Canis Major

- Canis Major incorporates multiple bright stars, including Sirius, within its night sky boundaries. This makes it readily identifiable in the night sky, particularly in the wintertime.

- Sirius is also the southern head of the asterism known as the "Winter Triangle."

- Canis Major is located due east of the prominent constellation Orion.

- Canis Major is one of the two hunter's dogs in Greek mythology. That is, Orion the Hunter did his thing with two faithful dogs at his side.

Crater the Cup

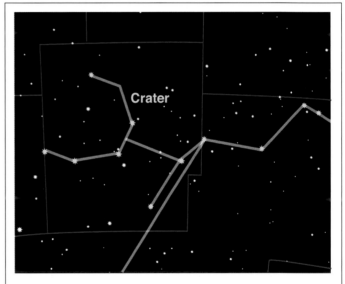

Crater

- Crater the Cup is an unremarkable constellation with a more remarkable place in mythology. Crater the Cup rests on the back of Hydra the Snake.

- Crater received its name from the Greek god Apollo's cup. When Apollo asked his pet crow, Corvus, to fetch him some water, the slothful bird dilly-dallied.

- Apollo punished his insubordinate pet by placing Corvus in the sky next to Crater—a cup that he could, for all eternity, only look at and never drink from.

Start by casting your eyes eastward at the constellation Orion the Hunter. This particular star pattern actually creates a "hunter" outline. With the easily spotted three stars in a neat row, locate the hunter's belt. Look then for the bright star Rigel at the hunter's right foot and Betelgeuse at the hunter's left shoulder. The Great nebula is near the middle of Orion's sword.

If you venture below Orion's belt of stars, you'll cross over into Canis Major the Big Dog, with the brightest star in the night sky, Sirius, alerting you of your space location.

Meanwhile, Taurus the Bull is the constellation that appears

to be charging the great hunter Orion. The bull "horns" shaped like a "V" are part of the Hyades open cluster of stars. And the bright star Aldebaran supplies Taurus's bull's eye.

A bit to the west of Taurus is the constellation Gemini the Twins, which is expressive in the winter night sky. This constellation looks more like a series of lines running parallel with one another. The two bright stars, Castor and Pollux, the "twins," will pinpoint your location and enable you to further explore the winter constellations.

Orion the Hunter

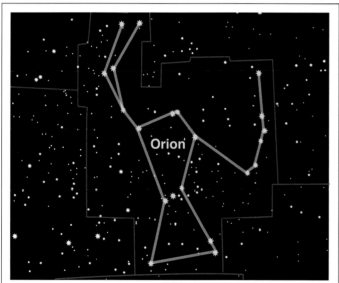

Sextans the Sextant

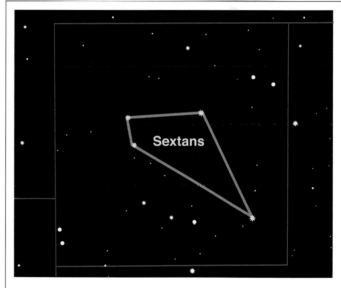

- Orion the Hunter is arguably the most famous constellation in the night sky.

- The constellation combines ease of location with patterns of stars that actually resemble a hunter with a crossbow.

- Orion also exhibits a readily spotted asterism of three bright stars in a row known as "Orion's Belt."

- The constellation is home to two of the brightest stars in the night sky, too: Rigel and Betelgeuse. The latter is a humongous red supergiant star in the twilight of its stellar life.

- Sextans the Sextant is one of the forty modern constellations.

- You can locate its border just below Regulus, the bright foot star in Leo the Lion.

- Sextans, the area of sky below Leo, is indisputably dark— a seemingly empty patch of night sky.

- But although not a prominent constellation, Sextans harbors myriad deep-sky objects, including galaxies. The NGC 3115, also known as the "Spindle galaxy," is estimated at more than twice the size of our Milky Way.

137

SPRING CONSTELLATIONS

Prominent spring constellations include Bootes, Cancer, Corona Borealis, Hercules, and Ursa Major

Springtime in the Northern Hemisphere night skies is considered a transitional time—between the prime starwatching seasons of winter and summer. Nevertheless, spring brings a number of constellations that are at their peak positions, including Bootes, Cancer, Canes Venatici, Corona Borealis, Corvus, Crate, Hercules, Leo, Serpens, Ursa Major, and Virgo.

Ursa Major the Great Bear is a circumpolar constellation visible throughout the year. It is, however, best viewed during the spring months. Ursa Major is the third-largest constellation and one of the most easily found in the night sky courtesy of its famous asterisms the Big Dipper and the Small Dipper. Although many people assume they are, these star

Canes Venatici the Hunting Dogs

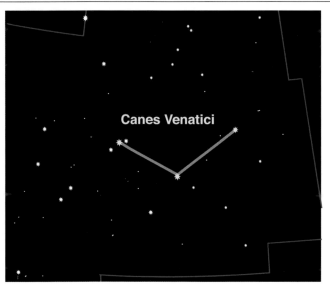

Corvus the Crow

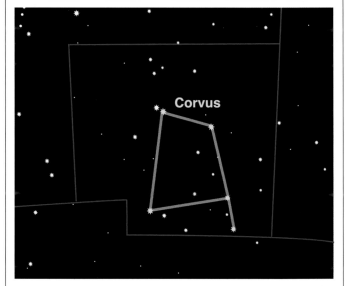

- Canes Venatici is nicknamed "the Hunting Dogs." It is a modern constellation that was at one time considered part of Bootes.

- Canes Venatici borders Bootes, Ursa Major, and Coma Berenices.

- The constellation can be spotted due south of the Big Dipper.

- Johannes Hevelius, one of the most prominent observational astronomers of the seventeenth century, is responsible for putting Canes Venatici and several other constellations on the celestial map.

- Corvus the Crow is a small constellation situated near Spica, the brightest star in Virgo. Cast your eyes immediately to the southwest of Spica, and you are in Corvus.

- Although it is not sprawling in size, Corvus is a worthwhile spring constellation to plow through.

- Within its compact boundaries are two colliding galaxies called the "Antennae galaxies," NGC 4038 and NGC 4039.

- To discern these galaxies, you will need clear, dark skies and a telescope with sufficient aperture.

patterns are not constellations in their own right. *Asterism* is the term applied to these smaller patterns of stars. Asterisms can be in specific constellations—like the Big Dipper in Ursa Major—or spread over one or more constellations.

Job one is finding the Big Dipper in the night sky with its distinctive bowl shape housing two "pointer" stars. These stars identify where Polaris, the North Star, is located. By employing Polaris as the ultimate stargazing point of reference, you can further star hop across the night sky. For now, stick with the Big Dipper and travel the celestial equivalent of downhill beginning at its bowl until you locate the bright star Regulus, which puts you in the middle of Leo the Lion. By letting your eyes dance across the extensive arc of the Big Dipper's handle, you will eventually encounter another vivid star, Arcturus in Bootes the Herdsman.

While exploring the spring constellations, don't ignore Hercules the Hero with its two bright stars. Springtime is also ripe for checking out Cancer the Crab, which accommodates the Beehive cluster, M44, a popular stargazing attraction.

Serpens Caput the Snake's Head

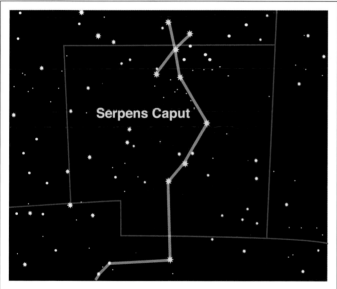

- Serpens the Serpent has the distinction of being a bifurcated constellation with two separate parts: Serpens Caput (the Snake's Head) and Serpens Cauda (the Snake's Tail).

- Although it has two noncontiguous halves, Serpens is classified as one constellation.

- Ophiuchus the Serpent Holder intersects the constellation.

- In mythology, Ophiuchus is the serpent bearer, holding the snake's head and tail on each side of him.

Ursa Major the Great Bear

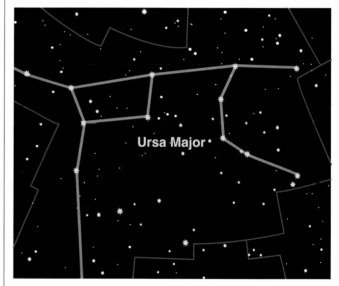

- Ursa Major the Great Bear is actually discernible throughout much of the year in the Northern Hemisphere. Spring is nonetheless a fine time to survey its residents.

- The Big Dipper is the constellation's most famous tenant.

- The Big Dipper is called the "Plough" in the United Kingdom.

- Ursa Major borders on multiple constellations: Draco, Camelopardalis, Lynx, Leo Minor, Leo, Coma Berenices, Canes Venatici, and Bootes.

SUMMER CONSTELLATIONS

Prominent summer constellations include Aquarius, Capricornus, Corona Borealis, Pegasus, and Sagittarius

The summer night skies in the Northern Hemisphere are teeming with visible and sharp constellations. Stargazers are supplied with prime views of summer constellations: Aquarius, Aquila, Capricornus, Cepheus, Corona Borealis, Cygnus, Delphinus, Lyra, Pegasus, Sagitta, Sagittarius, Scutum, and Vulpecula.

Foremost, check out the summer night skies and locate the Summer Triangle, which is an asterism with three vertices stars: Altair, Deneb, and Vega. These three stars represent the brightest in their respective constellations: Aquila, Cygnus, and Lyra. Indeed, this asterism touches three prominent summer constellations.

Vega, by the way, is the fifth-brightest star in the night sky and

Corona Borealis the Northern Crown

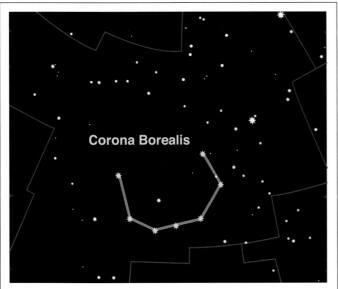

Corona Borealis

Sagitta the Arrow

Sagitta

- Corona Borealis the Northern Crown is a relatively small constellation in the northern sky.

- Corona Borealis features an alluring asterism in the shape of a semicircle. Alphecca, aka "Gemma," is the brightest star and the centerpiece of what is known as the "Northern Crown."

- The Northern Crown includes the curious variable star R Coronae Borealis, which periodically disappears from view when carbon condensation swaddles it in celestial soot.

- Sagitta the Arrow is a minuscule—the third-smallest, as a matter of fact—summer constellation. Although it is dim, it is easily found in the night sky.

- Sagitta, along with Vulpecula, lies within the Summer Triangle of stars. However, neither constellation contains a star that makes up the triangle.

- The Summer Triangle's three point stars are Deneb in Cygnus, Vega in Lyra, and Altair in Aquila.

- Sagitta signifies the arrow that Hercules used to earn his acclaim as a hero.

rates second in the northern celestial hemisphere. It is three times the size of our sun and appears bluish-white to stargazers. Deneb is actually twenty-five times the sun's dimensions and shines sixty thousand times brighter. It is, however, quite a distance a way. Nevertheless, Deneb is nicely visible in Cygnus the Swan, which also houses the famous Northern Cross.

The summertime also supplies you with a fine view of Sagittarius the Archer. This is an area in space that is packed with celestial objects and activity. Numerous nebulae can be found here, along with the Milky Way's galactic center.

Lyra the Harp

- Lyra the Harp can be spotted in the midsummer night sky by first finding the bright star Vega, which is almost directly overhead at this time of year.

- Vega is the brightest star in Lyra and one of the brightest in the night sky.

- Vega is one of the three point stars in the Summer Triangle.

- The constellation is shaped something like an asymmetrical rectangle—the harp strings—with Vega jutting slightly off from one of its corners.

Vulpecula the Fox

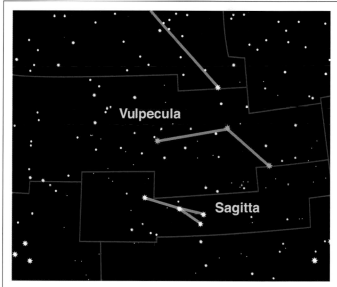

- Vulpecula the Fox is not a well-known constellation.

- It is nonetheless a summer constellation because of its location smack dab in the heart of the Summer Triangle, an asterism that countless stargazers check out during this season of the year.

- The Dumbbell nebula, M27, is a considerable and glowing planetary nebula in Vulpecula.

- Under optimal conditions—clear, dark sky—the Dumbbell nebula can be viewed with a good pair of binoculars. It appears as a gleaming disk of gas and dust.

FALL CONSTELLATIONS

Prominent fall constellations include Andromeda, Aries, Cassiopeia, Pegasus, and Perseus

When contrasted with night skies in the summer and winter, night skies in the Northern Hemisphere's fall are relatively quiet. Nonetheless, there are constellations that shine in the autumn months: Andromeda, Auriga, Camelopardalis, Cassiopeia, Cetus, Lepus, Pegasus, Perseus, Pisces, Sculptor, Taurus, and Triangulum.

Foremost, the autumn months are ideal times to check out the Great Square of Pegasus. Cast your eyes to the northeast close to the horizon. Here you will encounter a group of four stars with seemingly uniform brightness that form a neat square in outer space. The stars are Scheat, Alpheratz, Markab, and Algenib. But what's especially interesting about

Andromeda the Chained Maiden

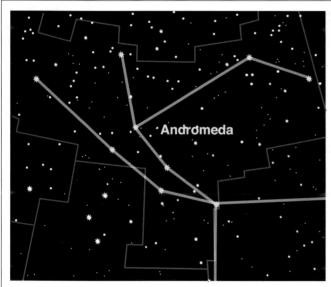

- The constellation Andromeda is oddly V-shaped and resides near the north celestial pole.

- It is nicely viewed in the autumn in the Northern Hemisphere, although it can also be seen throughout the summer.

- Andromeda's most famous tenant is our galactic neighbor.

- Look for the Andromeda galaxy about halfway up, and to the right, of the constellation's distinct V outline.

Lepus the Hare

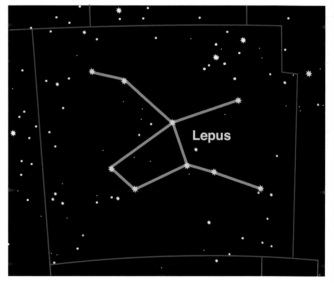

- Lepus the Hare is a small constellation hugging the horizon in autumn and winter. It can be found beneath the feet of Orion the Hunter.

- Mythology depicts a frightened hare trembling in the shadow of the mighty hunter.

- The globular cluster M79 can be observed in Lepus.

- This cluster is in an atypical location for these kinds of stellar fraternities. It is not the proverbial shouting distance from the Milky Way's galactic center but rather 60,000 light-years away.

this region of space is the apparent emptiness in the midst of the "square." From a naked eye's perspective, there is nothing there but darkness. This visual supplies the Great Square of Pegasus with a particular aura. In fact, there are faint stars and celestial objects within the square's imaginary boundaries, but they are difficult to see without enhancing equipment.

The constellation Andromeda the Chained Maiden is also nicely on display in the fall. This area of space is home to the Andromeda galaxy, M31, one of the more distant galaxies that can be observed with the naked eye.

ZOOM

Earth revolves around the sun in a counterclockwise direction. It takes one year to complete an orbit. And throughout this annual journey, the night skies are ever-changing portals into unique portions of the vast universe. Changing seasons ushered in and out by Earth's unflagging orbit introduce and bid farewell to constellations.

CONSTELLATIONS

Cetus the Whale

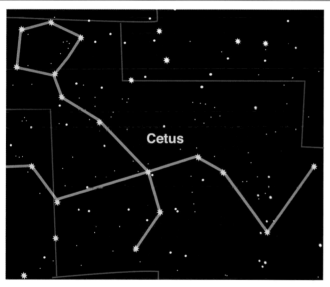

- Cetus the Whale is a northern sky constellation in an area of the sky associated with water and water-related things.

- Pisces the Fishes, Eridanus the River, and Aquarius the Water Bearer are among bordering constellations.

- Cetus houses Mira, the first variable star discovered.

- Although Cetus is not a zodiac constellation, the sun's ecliptic passes near its borders. Planetary visibility within Cetus occurs for short spans.

Perseus the Hero

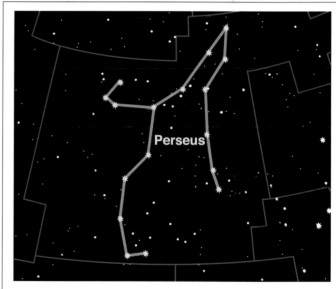

- The constellation Perseus the Hero is highly visible in the fall and winter.

- Many of the stars within this constellation are suffused in the warm glow of the Milky Way.

- Make it a point to check out Algol, Perseus's most

observed star. It's the "winking" star in the hero's right leg.

- Algol is actually a binary star with a fainter star orbiting it. When the fainter star passes in front of its brighter companion every few days, Algol dims—winks, if you will.

ARIES-TAURUS

Aries the Ram and Taurus the Bull are constellations with vastly different inhabitants

Courtesy of Earth's orbit around its parent star, our sun completes a full ring around our sky every year, an annual journey known as an "ecliptic." And as the sun follows along its ecliptic plane, it appears to pass through various constellations on its yearlong slog through the seasons. The constellations that host these ecliptic visitations are called the "zodiac

constellations." Whereas there are eighty-eight recognized constellations, there are just thirteen zodiac constellations.

Aries the Ram is a zodiac constellation. Situated between Pisces to its west and Taurus to its east, Aries is not particularly well known. This is because the constellation contains comparatively dim stars within it boundaries, the brightest

Aries the Ram

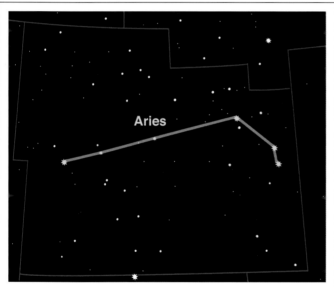

- Aries the Ram is both a zodiac constellation and, for horoscope devotees, the first astrological sign.

- The general star population in Aries is dim. But although it is not a stargazer's playground, it hosts its fair share of meteor showers.

- Annual meteor showers that occur in Aries include the daytime May Arietids and the Autumn Arietids.

- The constellations that border Aries are Perseus, Triangulum, Pisces, Cetus, and Taurus.

Taurus the Bull

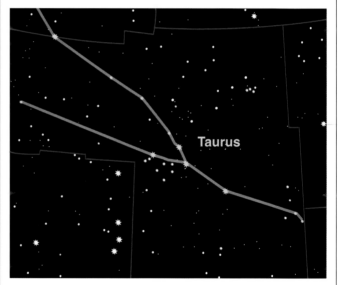

- The Taurus the Bull constellation is defined by its distinctive V shape of stars.

- This group of stars—an asterism—forms the celestial bull's face. The bright and orange star Aldebaran nobly plays the role of the bull's eye.

- Taurus borders on several constellations: Auriga, Perseus, Aries, Cetus, Eridanus, Orion, and Gemini.

- Aldebaran is near the sun's ecliptic.

being the not-especially-bright Alpha Arietis. The deep-sky objects in the constellation are likewise faint. Hence, Aries doesn't attract the attention of amateur stargazers.

Taurus the Bull is another zodiac constellation. Much more examined than Aries, Taurus accommodates the ultra-bright star Aldebaran within its borders. It is home to one of the best-known star clusters, the Pleiades, also known as the "Seven Sisters." This impressive celestial picture can be observed readily with the naked eye. With either a pair of binoculars or a telescope, you can detect many more stars.

MAKE IT EASY

The reason the zodiac constellations are important players in the overall science of astronomy has nothing to do with astrology and horoscopes. Rather, it's that the sun, moon, and planets in our solar system are always on or nearby the ecliptic. When starwatching, this astronomical reality should always be kept in mind.

Recognizable Taurus

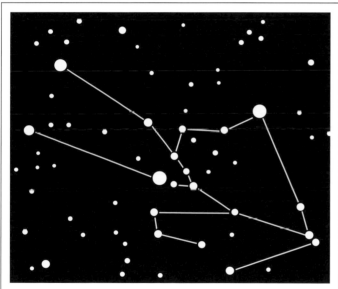

- Taurus the Bull is one of the more recognizable constellations.

- Taurus lies southwest of Orion. In Greek mythology, it was said that the belligerent Taurus huffed and puffed in the direction of the great hunter.

- The Hyades star cluster can be found in Taurus. Its brightest stars, coupled with Aldebaran, form the V-shaped bull's head.

- Aldebaran is not part of the cluster. The star is located much closer to Earth.

Zodiac Constellations

- Annually our sun appears to travel eastward and complete a full circle around the sky.

- This movement, known as the "ecliptic," is the consequence of Earth's orbit around the sun.

- The area of sky that host the ecliptic at varying points during the year are called the "zodiac constellations."

- Courtesy of astrology, there are twelve well-known zodiac constellations. However, a thirteenth exists: Ophiuchus.

GEMINI-CANCER

Gemini the Twins and Cancer the Crab accommodate diverse celestial bodies along the sun's ecliptic

Following on the heels of Aries and Taurus, the sun's ecliptic path takes it into the zodiac constellations of Gemini the Twins and then Cancer the Crab. Gemini is situated in the night sky between Taurus to its west and Cancer to its east. The constellations of Auriga and Lynx can be spotted to its north with Monoceros and Canis Minor to its south.

Gemini is, of course, most renowned for its twin stars, Castor and Pollux, which appear close together in the night sky and are the wind beneath the wings of the constellation's legendary nickname. Within its borders, Gemini houses the open cluster M35 and the Eskimo nebula (NGC 2392). The latter is sometimes referred to as the "Clown Face nebula"

Gemini the Twins

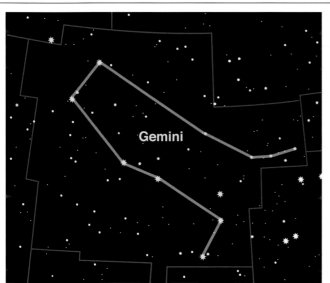

- Gemini the Twins is a zodiac constellation most known for its twin stars, Castor and Pollux.

- These bright stars appear quite close together, although in the celestial ether they are far apart.

- The stars' visible closeness is what inspired the mythology of the twins.

- Less-bright stars sloping downward from Castor and Pollux could—with a little imagination—resemble two human figures.

Gemini and the Naked Eye

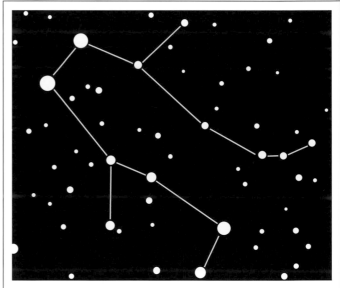

- Gemini is a constellation that can be observed with the unaided eye.

- Almost directly overhead in midwinter, its celestial territory can be seen just off to Orion's left.

- Gemini's borders touch multiple neighbors: Lynx, Auriga, Taurus, Orion, Monoceros, Canis Minor, and Cancer.

- Gemini is home to popular stargazing attractions, including M35, an open cluster of stars. And just to the southeast of M35 is another alluring open cluster known as "NGC 2158."

because it looks something like a human head sporting a parka-style hood. At 2,870 light-years away, this nebula can be seen and appreciated with most telescopes.

The constellation Cancer is not a very popular region of space for amateur stargazers. As far as constellations go, Cancer is small, and the stars on its limited stage are dim. It is, nonetheless, home to the Praesepe, M44, an open cluster of stars also known as the "Beehive cluster." Cancer is situated between Gemini to its west and Leo the Lion to its east.

Cancer the Crab

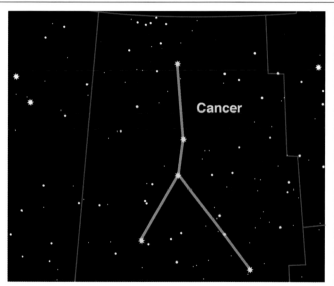

- Despite its notoriety as a zodiac constellation, Cancer the Crab is small with generally weak stars.

- 55 Cancri, which is actually two stars—a yellow dwarf and a red dwarf—resides in Cancer. Astronomers have identified multiple planets orbiting the bigger yellow dwarf star—55 Cancri A.

- This solar system, with extrasolar planets, is closer than any others seen.

- Cancer's most popular stargazing attraction is the Praesepe, M44, the nearest open cluster of stars to our solar system.

Dim Cancer

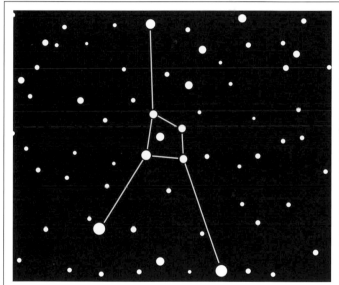

- Although Cancer is a dim constellation, it is bordered by two brighter zodiac constellations: Gemini and Leo.

- There are star formations in Cancer that are supposed to resemble a crustacean. If you look hard enough, you just might be able to make out two claws.

- Next to the crab's head is the famous Praesepe, aka "Beehive cluster." To the naked eye the cluster appears as a hazy cloud.

- Use binoculars or a telescope for a more intimate look at this bright star cluster.

LEO-VIRGO

Leo the Lion and Virgo the Maiden are areas of space brimming with bright galaxies

Leo the Lion and Virgo the Maiden are zodiac constellations with a heaping helping of diverse celestial bodies within their space boundaries. In Leo, the bright star Regulus can be effortlessly detected in the night sky, along with additional bright stars Denebola and Leonis.

Leo contains multiple galaxies within its boundaries, including M65, M66, and NGC 3628, which comprise what is known as the "Leo Triplet." This grouping can be observed with most levels of telescopes. Both M65 and M66 are spiral galaxies and similar to the Milky Way in structure. M66 contains a bevy of luminescent star clusters and extensive spiral arms. Additional bright galaxies in the Leo constellation

Leo the Lion

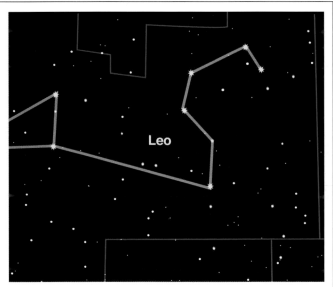

Leo

- Leo the Lion is luminescent.

- Regulus, one of the brightest stars in the night sky, assumes the position as the celestial lion's tail. Denebola and Leonis are nearby stars of standing.

- Leo is a large constellation with many neighbors

touching its borders: Ursa Major, Leo Minor, Lynx, Cancer, Hydra, Sextans, Crater, Virgo, and Coma Berenices.

- One of the nearest stars to Earth, Wolf 359, calls Leo home. Wolf 359 is a red dwarf star. You will need a large telescope to find it.

What Is Astrology?

- Astrology studies the positioning of the sun, planets, and the moon.

- In astrology, the locations and movements of these celestial bodies are presumed to influence human behavior and events.

- Astrology has put the zodiac constellations on the map because of the sun's ecliptic, which passes through specific areas of the sky.

- Astrology is extraordinarily popular in cultures throughout the world and is a cottage industry.

include M95 and M96. But Leo is hardly done yet with sights for stargazers' eyes. There is the Leo Ring. This primordial cloud of gases—hydrogen and helium—is thought to be a remnant of the Big Bang and the universe's formation.

Virgo is just to the west of Leo and next in line along the sun's ecliptic. The zodiac constellation of Libra is to its east. With its sprawling expanse, Virgo is the second-largest constellation. Virgo covers a lot of space and can be readily located in the sky thanks to its brightest star, Spica. Virgo is chock full of diverse galaxies, including M49, M58, and M84.

M49 is an elliptical galaxy; M58, a spiral galaxy; and M84 an atypical galaxy sometimes defined as "lenticular."

The constellation Virgo is also home to the Sombrero galaxy, M104. This galaxy is a different kind of spiral galaxy and shaped like—you guessed it—a sombrero. Finally, within Virgo's rambling space neighborhood are extra planetary systems of stars, including 70 Virginis and 61 Virginis. These stars, with planets orbiting them akin to the eight planets orbiting our sun, fascinate astronomers and open up a wealth of possibilities as to extraterrestrial life beyond Earth.

Virgo the Maiden

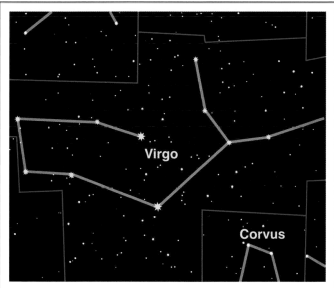

- Virgo the Maiden is one of the more happening of all zodiac constellations.

- Its brightest star, Spica, makes locating Virgo in the night sky a relatively simple task. The arc of the Big Dipper leads you to first Arcturus in the constellation Bootes and then Spica.

- Virgo's boundaries include one of the points where the celestial equator intersects with the ecliptic.

- Virgo has the distinction of accommodating more known extrasolar planets than does any other constellation.

Rambling Virgo

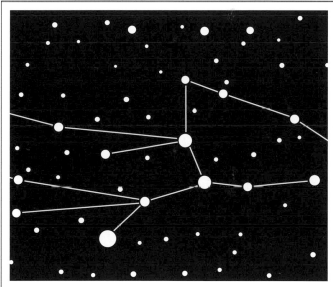

- Virgo's borders touch numerous constellations: Bootes, Coma Berenices, Leo, Crater, Corvus, Hydra, Libra, and Serpens Caput.

- Eleven Messier's objects are found within the constellation's boundaries.

- These Messier's objects are mostly galaxies found in what is called the "Virgo cluster." This sprawling intergalactic neighborhood, which crosses over into the constellation Coma Berenices, is immense, with approximately two thousand residents.

LIBRA-SCORPIUS

Libra the Scales and Scorpius the Scorpion supply contrasting pictures of zodiac constellations

Libra the Scales is just to the west of Virgo and the sun's next visitation during its ecliptic run through the zodiac constellations. Scorpius borders Libra to its west.

Libra is a distinctly more placid constellation than Virgo on its one side and Scorpius on its other. For instance, its territorial boundaries in space contain no Messier's objects. But as is

always the case on the celestial frontier, in each space neighborhood, something is always going on. Sometimes we cannot see it with our naked eyes or even with binoculars or a telescope. Sometimes it takes advances in technology to make discoveries of import. This occurred in Libra with the star Gliese 581. This not-especially-unique star is parent of a planetary

Libra the Scales

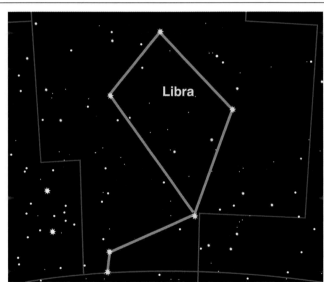

- Libra the Scales is a zodiac constellation with no first-magnitude stars to its credit. Hence, it is not a bright region of night sky.

- Compared with its zodiac neighbors, Virgo and Scorpius, Libra is a rather extensive patch of celestial nothingness.

- Libra is the only zodiac constellation depicted by a non-living creature.

- Libra's mythological underpinning is scales held by the goddess of justice, Astaea.

Sleepy Libra

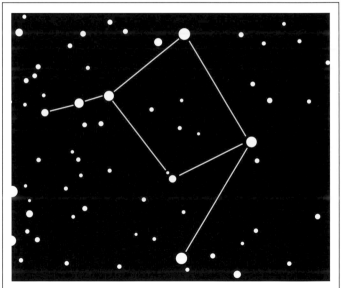

- The brightest stars in Libra form what looks like a quadrangle.

- Libra's brightest star is Beta Librae, aka "Zuben-eschamali." To naked-eye observation the star appears greenish.

- You can locate Beta Librae

a fair distance to the northwest of the bright star Antares in Scorpius.

- Libra's balanced scales were associated with this patch of sky because the sun's ecliptic once brought it through the constellation on the vernal equinox—equal amounts of day and night.

system containing a minimum of four planets orbiting around it. Such astronomical findings are always exhilarating because they indicate that our solar system is not wholly unique in the vastness of space. Libra has not only two zodiac constellations bordering it on its east side and west side but also additional constellations in different areas. Serpens, Hydra, Centaurus, Lupus, and Ophiuchus touch Libra's boundary lines. Getting to know where the constellations are vis-à-vis one another makes stargazing more rewarding.

Scorpius, west of Libra, is a wholly different zodiac constellation. It spans an extensive area of space in the Southern Hemisphere. Scorpius is the region of the cosmos where our parent galaxy's galactic center shines bright. Indeed, this is a well-lit area of the celestial beyond because of the galaxy's luminescent and scorching core.

Scorpius has multiple bright stars, notably Antares, and various deep-sky objects of note. The deep-sky objects worth checking out include the Butterfly cluster, M6, and the Ptolemy cluster, M7. Globular clusters M4 and M80 also paint compelling star portraits in Scorpius's snippet of the night sky.

Scorpius the Scorpion

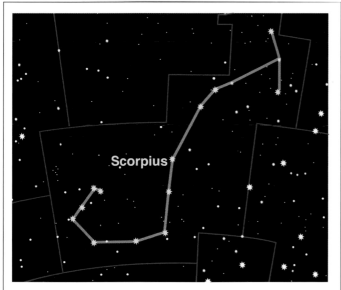

- From amateur astronomers' perspectives, Scorpius the Scorpion is among the favorite zodiac constellations.

- Located in the southern sky, its bright stars resemble a celestial arachnid. Antares, the red supergiant star, is easily distinguishable at the heart of the creature.

- Scorpius's position on the Milky Way, riding high in the southern sky or hugging the horizon in the north, furnishes you a view of many deep-sky objects.

- Globular cluster M4, which is one of the closest and most impressive clusters of stars, is found in Scorpius.

Scorpius in the Milky Way Sheen

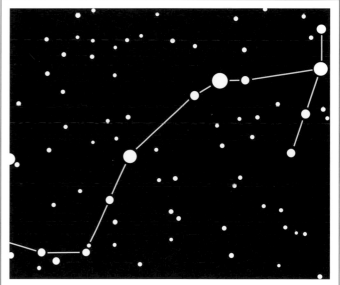

- Scorpius the Scorpion is both a large and recognizable patch of night sky.

- Neighboring constellations include Sagittarius, Ophiuchus, Libra, Lupus, Norma, Ara, and Corona Australis.

- Based on its coordinates in the night sky, the constellation greatly benefits from the gentle sheen of the Milky Way.

- An exception to the rule in the eighty-eight official constellations, Scorpius the Scorpion's star formations actually look like what they are supposed to look like.

OPHIUCHUS-SAGITTARIUS

Unlike Sagittarius the Archer, Ophiuchus the Serpent Bearer is a frequently overlooked zodiac constellation

Ophiuchus the Serpent Bearer is a large constellation of stars in the environs of the celestial equator. The sun's ecliptic saunter across the sky brings it through Ophiuchus in December. This fact makes it a zodiac constellation, but it is nonetheless overlooked because of the non-science of astrology, which does not count Ophiuchus among its twelve zodiac signs.

Conversely, Sagittarius the Archer is a well-known zodiac constellation that is associated with the month of December. Ophiuchus borders Sagittarius and Serpens to its east. Libra and Scorpius stand guard on its southern border. Ophiuchus is northwest of the Milky Way's bright nucleus. It is the home of Barnard's Star, which is among the closest stars to Earth.

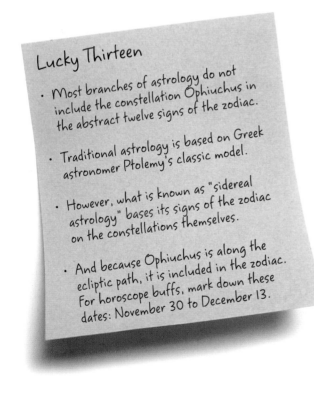

Lucky Thirteen

- Most branches of astrology do not include the constellation Ophiuchus in the abstract twelve signs of the zodiac.

- Traditional astrology is based on Greek astronomer Ptolemy's classic model.

- However, what is known as "sidereal astrology" bases its signs of the zodiac on the constellations themselves.

- And because Ophiuchus is along the ecliptic path, it is included in the zodiac. For horoscope buffs, mark down these dates: November 30 to December 13.

Ophiuchus the Serpent Bearer

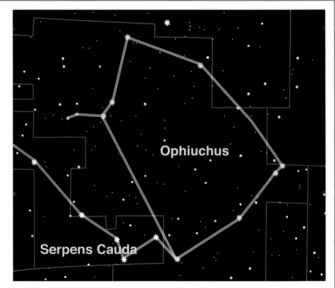

- Ophiuchus the Serpent Bearer is a rather expansive but little-explored constellation.

- It is near the celestial equator and one of the largest in this region of the sky.

- Ophiuchus accommodates seven deep-sky Messier's

objects. The individual objects are globular clusters: M9, M10, M12, M14, M19, M62, and M107.

- The proliferation of these clusters of stars is reason enough to check out this frequently bypassed constellation.

But although this star is nearby, it is not visible to the naked eye. However, with a basic telescope, you can locate this red dwarf star. Sagittarius is on the stellar doorstep of the Milky Way's epicenter. A compelling view of this dense area of our parent galaxy is seen within Sagittarius's borders, which are brimming with star clusters, like M55, and nebulae, like the Lagoon nebula, M8, and the Omega nebula, M17. Sagittarius has a fair number of constellations that touch its boundaries at various points, including Ophiuchus, Scorpius, and Capricornus.

Sagittarius the Archer

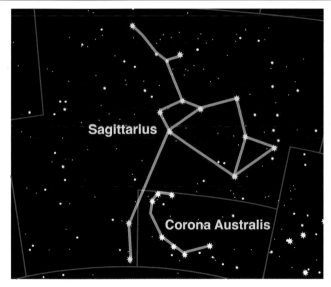

- As viewed from our earthly perspective, the constellation Sagittarius is home to the brightest stretch of the Milky Way—its galactic center.

- Accommodating the Milky Way's core within its boundaries ensures that Sagittarius the Archer is a

  night-sky hotspot.

- The Milky Way's tender glow guarantees that Sagittarius is replete with deep-sky objects such as nebulae and star clusters.

- The bright and bulky globular cluster known as "M55" is a favorite stargazing target.

Sagittarius Summertime

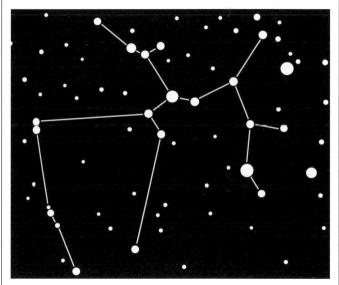

- If you are endeavoring to decipher the contours of an archer among Sagittarius, you could conceivably make out a rather irregular, stick-like figure drawing his bow.

- Sagittarius's fertile, creamy-looking star fields are the byproduct of its nearness to the Milky Way's nucleus.

- The constellation Sagittarius has more individual stars with known planets orbiting them than any others.

- With the ecliptic taking the sun through Sagittarius in the early wintertime, this teeming constellation is best observed in the summer.

CAPRICORNUS-AQUARIUS-PISCES

These constellations lie in a region of the sky known as the "Sea"

The final three zodiac constellations are Capricornus the Goat, Aquarius the Water Bearer, and Pisces the Fishes. Each one of these constellations calls home a section of the night sky called the "Sea" or "Water." These appellations derive from the prominence of named constellations in the area associated with water or watery things, like Cetus the Whale and Eridanus the River.

Capricornus is a southern sky constellation and a rather weak one at that. Among the zodiac constellations, only Cancer is fainter to observers. The only Messier's object within its boundaries is M30.

Aquarius, which lies between Capricornus to its west and Pisces to its east, is patently more visible than its ecliptic predecessor. Astronomers have located five stars within the constellation hosting planetary systems. Aquarius is also home to globular clusters M2 and M72 as well as open cluster M73.

Capricornus the Goat

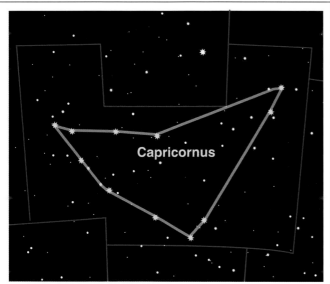

- Capricornus the Goat is the celestial next-door neighbor of Sagittarius. However, it doesn't profit from its proximity to the Milky Way's galactic center.

- Capricornus is among the faintest constellations in the night sky.

- Although it is the antithesis of a stargazer's paradise, it maintains one of the oldest mythological connections.

- Its depiction as a cross between a goat and a fish goes back to the Bronze Age.

Aquarius the Water Bearer

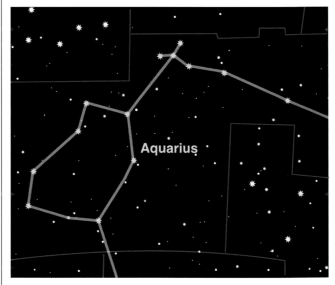

- Aquarius the Water Bearer is a dim zodiac constellation with nonetheless three globular clusters within its celestial borders.

- Aquarius also is home to the Saturn nebula NGC 7009, which resembles the planet Saturn through a telescope.

- The brighter nebula in the constellation is the Helix nebula NGC 7293.

- The mythological figure at the foundation of this constellation is a man pouring water from an urn, jar, or bucket.

Planetary nebulae on its stellar turf include the Saturn nebula (NGC 7009) and the Helix nebula (NGC 7293).

Pisces is yet another one of the faint zodiac constellations. Bordering with Aquarius to its west and Aries to its east, Pisces can be found in the equatorial region of the night sky. It is home to the spiral galaxy M74. There's really not a whole lot for stargazers to see with the naked eye here. Binoculars and telescopes supply the best picture of Pisces.

Faint Aquarius

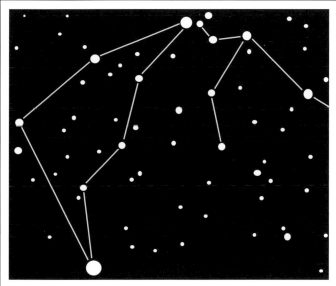

- Although Aquarius the Water Bearer is a generally faint celestial backdrop, it houses several deep-sky objects of interest to stargazers.

- The globular cluster M2 can be spotted with the naked eye.

- And the Helix nebula NGC

7293 is the closest planetary nebula, at a mere 400 light-years away, to Earth.

- Although some imaginative starwatchers claim to make out the shape of a boy, trying to decipher the contours of a human figure among the stars is typically an exercise in futility.

Pisces the Fishes

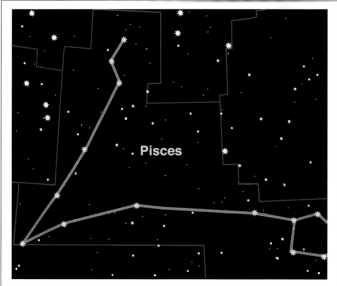

- Pisces the Fishes—or "Fish," as it is more commonly known—is a zodiac constellation of some renown. Pisces does not have any particularly bright stars but rather sports a conspicuous circlet of stars at the head of its western fish.

- Pisces also hosts the vernal equinox, the location in the sky where the sun intersects the celestial equator.

- Its one Messier's object is a spiral galaxy known as "M74."

DEEP-SKY OBJECTS

Deep-sky objects are astronomical bodies beyond our solar system

A popular pursuit of amateur astronomers is locating deep-sky objects. That is, celestial bodies beyond the sun, moon, and planets Mercury, Venus, Mars, Jupiter, Saturn, Neptune, and Uranus. Examples of deep-sky objects include open clusters of stars, globular clusters of stars, planetary nebulae, diffuse nebulae, dark nebulae, and galaxies beyond the Milky Way.

Ever-increasing technology has enabled previously unseen space bodies to materialize on the celestial sphere.

Nevertheless, ancient civilizations had their naked eyes on many deep-sky objects, including the Pleiades and Hyades star clusters in the zodiac constellation Taurus. The Andromeda galaxy was spotted and documented as far back as A.D. 905. In the Southern Hemisphere, stargazers have long known about the Large Magellanic Cloud (LMC) and the Small Magellanic Cloud (SMC). When conditions are right, these deep-sky objects are visible to the naked eye.

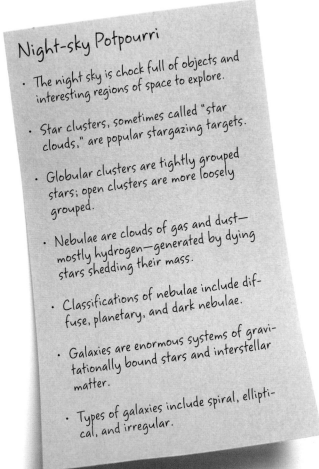

Night-sky Potpourri

- The night sky is chock full of objects and interesting regions of space to explore.

- Star clusters, sometimes called "star clouds," are popular stargazing targets.

- Globular clusters are tightly grouped stars; open clusters are more loosely grouped.

- Nebulae are clouds of gas and dust—mostly hydrogen—generated by dying stars shedding their mass.

- Classifications of nebulae include diffuse, planetary, and dark nebulae.

- Galaxies are enormous systems of gravitationally bound stars and interstellar matter.

- Types of galaxies include spiral, elliptical, and irregular.

Quintuplet Star Cluster

- The Quintuplet star cluster exists in the neighborhood of the Milky Way's galactic center.

- It is a dense celestial milieu teeming with young and massive stars, including the Pistol star.

- The Pistol star is among the most luminescent in the entire Milky Way but largely obscured by its own emissions of gas and dust known as the "Pistol nebula."

- This intriguing star cluster is approximately 100 light-years away from our galaxy's core.

Indeed, long before the invention of the telescope, people were detecting unusual objects in deep space. This shadowy quality makes deep-sky objects exceptional stargazing targets. Many can be seen in various constellations sans enhancing gear. Nevertheless, deep-sky objects naturally benefit from a closer look, and thus binoculars and telescopes can take naked-eye sightings to a more intimate level and bring them further to life. And, naturally, many deep-sky objects can be viewed only with telescopes.

ZOOM

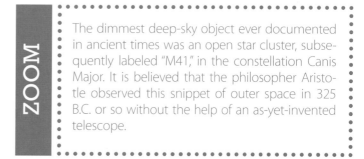

The dimmest deep-sky object ever documented in ancient times was an open star cluster, subsequently labeled "M41," in the constellation Canis Major. It is believed that the philosopher Aristotle observed this snippet of outer space in 325 B.C. or so without the help of an as-yet-invented telescope.

Nebula RCW 49

- The RCW 49 nebula has been dubbed a "stellar nursery": a dark, dusty, and gaseous environment perfect for star formation.

- With this nebula's center of older blue stars clustered together, NASA's Spitzer space telescope offers a rare infrared view.

- Newborn stars are interspersed amidst the expansive cloud. Gas streaks are visible in green and dusty wisps seen in pink.

- RCW 49, at 13,700 light-years away, can be found in the constellation Centaurus.

Hoag's Object

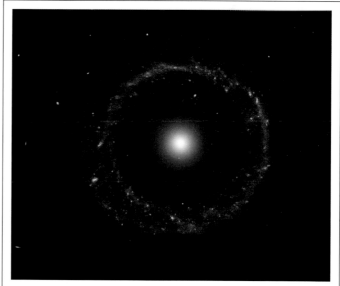

- Hoag's object is one of the more unusual galaxy discoveries. It is an irregular galaxy—not spiral or elliptical—known as a "ring galaxy."

- Its outside ring is replete with radiant blue stars, whereas nearer the center are red stars. The red stars are likely the older stars.

- Some astronomers speculate these ring-shaped galaxies form via the mergering of a small one with a larger one.

- Hoag's object is 100,000 light-years in diameter.

MESSIER'S OBJECTS

French astronomer's catalog of deep-sky objects includes star clusters, nebulae, and galaxies

If you are interested in locating deep-sky objects, it is imperative that you pay homage to Charles Messier. This French astronomer is the undisputed father of the deep sky. Sprinkled throughout books on astronomy, including this one, are myriad references to Messier's objects. They are often seen as just an "M" with a corresponding number attached. Messier cataloged his findings by number.

Charles Messier was born in 1730. His passion involved comets, and he habitually felt frustrated by outer-space false alarms. That is, by deep-sky objects that weren't what he hoped they were—comets—but rather star clusters, nebulae, and galaxies instead. He began systematically cataloging these deep-sky

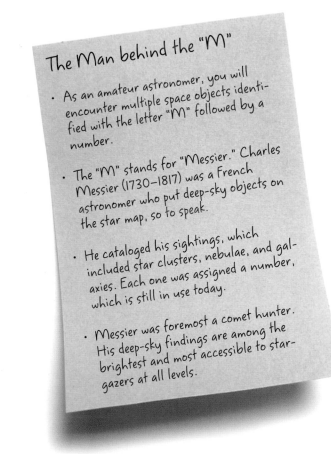

The Man behind the "M"

- As an amateur astronomer, you will encounter multiple space objects identified with the letter "M" followed by a number.

- The "M" stands for "Messier." Charles Messier (1730–1817) was a French astronomer who put deep-sky objects on the star map, so to speak.

- He cataloged his sightings, which included star clusters, nebulae, and galaxies. Each one was assigned a number, which is still in use today.

- Messier was foremost a comet hunter. His deep-sky findings are among the brightest and most accessible to stargazers at all levels.

Omega Nebula

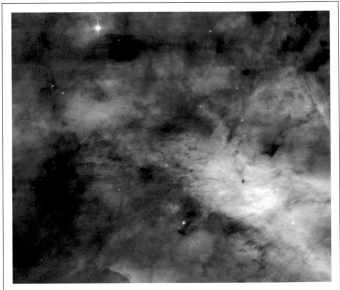

- The Omega nebula, M17, is known as an "emission nebula."

- The colors of the Omega nebula reveal star formation and the megaenergy generated in the process.

- Astronomers are confident that star formation in this nebula is ongoing, or only recently ended, because of its countless young and bright stars largely obscured by dust.

- The Omega nebula is sometimes referred to as the "Horseshoe nebula" or "Swan nebula."

sightings, assigning them numbers in the process.

Messier's original catalog contained forty-five deep-sky objects—M1 through M45. His total recordings eventually reached 103. Subsequently, latter-day astronomers added to the catalog seven deep-sky objects, which they noted Messier had observed, bringing the total to 110, where it stands today.

Based on Messier's European location, his deep-sky objects preclude portions of the Southern Hemisphere that never reveal themselves in the Northern Hemisphere. He thus never recorded the Large Magellanic Cloud and the Small Magellanic Cloud, seen only in Southern Hemisphere locales. And although these space bodies are never visible up north, they nonetheless meet the deep-sky object criteria established by Messier.

Although many of Messier's objects are visible with the naked eye, every one of the 110 chronicled in his catalog can be seen with binoculars and basic telescopes. No Hubble space telescopes are required. This stargazing reality is why Messier's objects are often the underpinning of amateur astronomy outings. Messier's objects span a wide range of celestial bodies in a wide range of places.

Wild Duck Cluster

- The open cluster M11 is more popularly called the "Wild Duck cluster." It is home to an estimated 2,900 stars.

- This extraordinarily dense star cluster houses multiple hundreds of first-magnitude stars. From our perspective here on Earth, first-magnitude stars are the brightest stars.

- The Wild Duck cluster is located in the constellation Scutum and can be observed with binoculars.

- Telescope users have likened M11's shape of dense stars to wild ducks in flight.

Messier 49

- The galaxy M49 is among the brightest inhabitants within the famous Virgo cluster of galaxies.

- It is an "elliptical galaxy," the designation affixed to galaxies that have oval contours and lack spiral arms.

- Spiral galaxies, like our Milky Way, are the most commonly observed galaxies.

- M49 is also known as "NGC 4472." Many celestial objects wear this "NGC" tag, which sometimes overlaps the Messier brand. "NGC" is the acronym for "New General Catalogue," another widely used cataloging.

ANDROMEDA GALAXY

The Andromeda galaxy is believed to be similar to our parent galaxy, the Milky Way

The Andromeda galaxy is located in the constellation Andromeda—yes, where it got its name. It is one of the most famous of Messier's deep-sky objects and is alternatively known as "M31." Plain and simply, the Andromeda galaxy is our Milky Way's next-door neighbor. That is, it's the nearest substantial galaxy to our own. In fact, the Andromeda galaxy is among the so-called Local group of galaxies, which includes M32 and M110, two exceptionally bright dwarf elliptical galaxies.

But the Andromeda galaxy is of particular interest to scientists and amateur stargazers alike because of its similarities to our parent galaxy. Both the Andromeda and the Milky Way are spiral galaxies containing hundreds of billions of

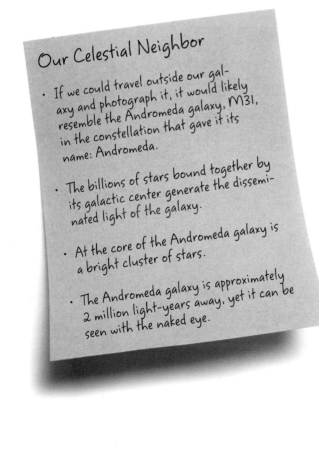

Our Celestial Neighbor

- If we could travel outside our galaxy and photograph it, it would likely resemble the Andromeda galaxy, M31, in the constellation that gave it its name: Andromeda.

- The billions of stars bound together by its galactic center generate the disseminated light of the galaxy.

- At the core of the Andromeda galaxy is a bright cluster of stars.

- The Andromeda galaxy is approximately 2 million light-years away, yet it can be seen with the naked eye.

Naked-Eye Visibility

- A naked-eye perspective of the Andromeda galaxy furnishes you with an illuminated blur.

- Binoculars can bring the image closer to you. However, telescopic viewings are recommended for viewing our galactic neighbor. With this magnified view, you can appreciate the galaxy's defined disk shape.

- Autumn is a good time to check out the Andromeda galaxy in the constellation of the same name.

- The galaxy is centrally located within Andromeda.

stars bound together by persuasive gravitational forces. Our neighboring galaxy is approximately 2.2 million light-years away. And its overall diameter is believed to be at least one and one-half times bigger than that of the Milky Way.

Just like the Milky Way, the Andromeda galaxy orbits around a central core—or bulge—where a colossal black hole is presumed to lurk. This mysterious makeup is quite commonplace in spiral galaxies like the Andromeda, Milky Way, and others with substantial mass. Andromeda's compelling galactic center pulls slighter galaxies into its orbit.

Tightly Wound Galaxy

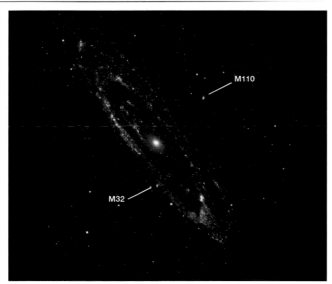

- The Andromeda galaxy's tightly wound, dramatic persona is a celestial sight for sore eyes.

- Two tiny elliptical galaxies can be spotted. These considerably smaller conglomerations of stars are satellites of the dominant Andromeda.

- X-ray discharges are visible in the regions of incredibly high energy.

- If you are having difficulty locating Andromeda, find Polaris, the North Star. Trace an imaginary line from the pole star through the "W" in Cassiopeia, and you'll be at the galaxy's doorstep.

Andromeda in Ultraviolet Light

- The ultraviolet image of the Andromeda galaxy reveals its distinct spiral arms. Noticeable tufts of blue indicate the presence of young and massive stars.

- The bright center of the galaxy showcases a dense region of much cooler and older stars.

- Courtesy of its extraordinary brightness and relative closeness, the Andromeda galaxy is one of only three galaxies that can be seen with the unaided eye.

- The Andromeda galaxy is huge: 260,000 light-years across.

DOUBLE CLUSTER IN PERSEUS
The Double Cluster of stars in the constellation Perseus is a celestial double feature

The well-known Double Cluster in the northern constellation of Perseus is approximately 7,000 light-years away from Earth. This twin cluster of stars (NGC 869 and NGC 884) can, in fact, be spotted with the naked eye under optimal conditions. Nevertheless, this impressive double cluster of stars is best viewed with a pair of binoculars or a telescope, allowing

you a much sharper look at the hundreds of bright stars in these clusters.

The Double Cluster in Perseus is regarded as one of the most striking deep-sky object finds not cataloged by French astronomer Charles Messier. Why did he miss the Double Cluster? Because he was foremost interested in comet

Two For the Price of One

- The Double Cluster in Perseus is two nearby open clusters of stars, NGC 884 and NGC 869.

- As star clusters go, the Double Cluster is considered positively youthful. NGC 884 is estimated at 3.2 million years of age. NGC 869 is approximately 5.6

million years of age.

- Compare these adolescent stellar ages with that of the famous Pleiades cluster, which is presumed to be one hundred million years old.

- The Double Cluster can be seen nicely with a pair of binoculars.

A Naked-Eye Perspective

- Under optimal stargazing conditions, the Double Cluster is observable with the naked eye. It appears as two illuminated blurs—quite close together—in the constellation Perseus.

- To find the Double Cluster in the night sky, look to the northernmost region of

Perseus, not too far from its border with Cassiopeia.

- The fall and winter are the best time to explore the Double Cluster.

- Both clusters are approximately 7,000 light-years away from Earth.

sightings, it is surmised that he purposely bypassed dazzling star clusters. Messier likely reasoned that they couldn't possibly be confused with comets.

It should be noted that the Double Cluster in Perseus is circumpolar, which means that it never rises or sets, as it were. That is, the twin clusters are visible above the horizon every night of every day of the year. But, as is always the case in the night sky, not all viewing times are created equal. Suffice it to say, it is best to check out the clusters when they are high in the sky and not hugging the horizon.

ZOOM

Many deep-sky objects are identified with the acronym "NGC." "NGC" stands for "New General Catalogue," which acknowledged 7,840 celestial bodies in the night sky. Compiled in the 1880s by J. L. E. Dreyer, who extrapolated the observations of astronomer William Herschel and his son, John, this extensive catalog includes nebulae, open clusters and globular clusters.

Single Open Cluster

- Open clusters are very common in the Milky Way. More than one thousand are known for certain to exist.

- It is suspected that there are many more star clusters than those already observed. Open clusters are found in large numbers in the galaxy's spiral arms.

- What makes the Double Cluster so unique is its dual nature. The clusters in Perseus are separated by approximately 300 to 400 light-years, which is the celestial equivalent of walking distance.

Ancient Find

- The ancient Greek astronomer Hipparchus recorded his observation of the Double Cluster in 130 B.C.

- There are several hundred stars in each cluster.

- Most of the stars in the Double Cluster are young and luminous blue stars.

- Sprinkled throughout the twin cluster is also a fair share of supergiant orange stars with incredible luster.

DEEP SKY: FALL-WINTER

ORION NEBULA

The Orion nebula is a colorful region of space where stars are born

The Orion nebula, M42, can be found in the constellation Orion. In fact, this diffuse nebula, as it is classified, is sometimes confused as a middle star in the Orion Belt. That is, Orion the Hunter's belt.

The Orion nebula, which occasionally is referred to as the "Great nebula in Orion" or "Great Orion nebula," is incontestably one of the more luminescent of deep-sky objects. It is therefore a popular and rewarding stargazing attraction. The

Orion nebula can, in fact, be spotted with the naked eye. However, without binoculars or a telescope, its nebulosity cannot be deciphered. This is why it appears as a star to so many casual observers of the night sky. But with visual aids, its gaseousness can be seen and fully appreciated. Indeed, the gases amidst the Orion nebula, believed to be mostly hydrogen, emit a radiant glow. The gases surround a mother lode of hot and young stars. This one-two punch of gaseous

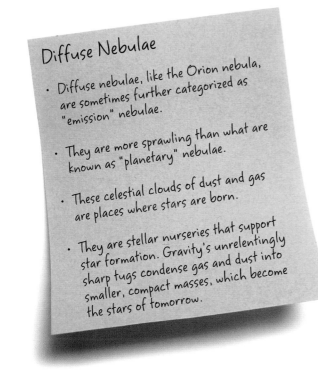

Diffuse Nebulae

- Diffuse nebulae, like the Orion nebula, are sometimes further categorized as "emission" nebulae.

- They are more sprawling than what are known as "planetary" nebulae.

- These celestial clouds of dust and gas are places where stars are born.

- They are stellar nurseries that support star formation. Gravity's unrelentingly sharp tugs condense gas and dust into smaller, compact masses, which become the stars of tomorrow.

Ultraviolet View

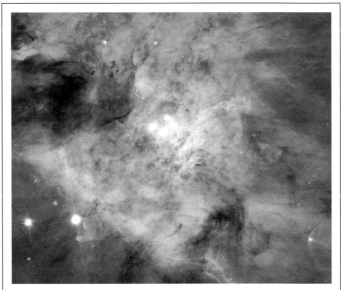

- The Orion nebula, M42, is located in the constellation Orion the Hunter. It is considered a diffuse nebula.

- This deep-sky object can be seen due south of the famous Orion belt of stars. The winter is a prime time to check out this multihued cloud of gas and dust.

- The Orion nebula is 24 light-years in diameter.

- The Trapezium is the open cluster of stars visible near the center of the nebula.

clouds and newly forming stars amidst the interstellar matter is a recipe for outer-space illumination. Indeed, the Orion nebula exudes a range of colors with areas of red and others of blue-violet. There is simultaneously a curious green shade that baffles astronomers as to its origins.

The Orion nebula is not only widely scrutinized by starwatchers but also picked apart by scientists who glean invaluable clues as to how stars are born in the heart of crumpling clouds of gas and space dust.

Enormous young stars and columns of incredibly dense gas fundamentally shape the Orion nebula. It is at once an intriguing visual but also the nearest interstellar region with active and evident star formation. In other words, this deep-sky object, which was first spotted with a telescope in 1610, is a veritable space laboratory and stellar nursery.

Inside a Stellar Nursery

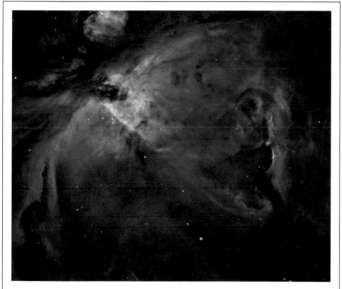

- The Hubble space telescope image of the Orion nebula is dramatic, supplying an intimate peek into a dynamic stellar nursery.

- The Orion nebula spans a considerable star-breeding area.

- At a relatively close 1,500 light-years from Earth, it furnishes us with a rare, close-up window into stellar evolution.

- This view unmistakably reveals how the nebula's hottest and brightest stars are penetrating the otherwise-obscuring dust and gas.

The Passion of Youth

- NASA's Chandra X-ray Observatory has surveyed the Orion nebula as never before, revealing 1,400 young stars similar to our sun in its infancy.

- These sunlike stars are generating flares that overwhelm in magnitude and frequency anything coming off the present-day sun.

- Scientists believe we are witnessing the way it was, the way our sun behaved in its adolescence.

- All across the Orion nebula, scientists are investigating stars one to ten million years old.

PLEIADES CLUSTER

This tightly knit cluster of stars contains extremely young and bright stars

The Pleiades cluster, M45, also known as the "Seven Sisters" of Greek mythology, is a dazzling open cluster of stars that can be found in the constellation Taurus the Bull.

And although the cluster contains hundreds of stars, only fourteen can be seen with the naked eye and then only under optimum stargazing conditions, including the right time of year. The Pleiades cluster is estimated to have appeared on the space scene some one hundred million years ago, which makes the stars in the assemblage an extremely young bunch. By comparison, our sun is more than four billion years old. The Pleiades cluster is 425 light-years from Earth, which explains why some of the extraordinarily bright stars amidst

Pleiades Cluster in Infrared

- The Pleiades star cluster, M45, is believed to have formed about one hundred million years ago, making the stars therein approximately one-fiftieth the age of our sun.

- Although the majority of the approximately three thousand stars are not perceptible to the naked eye, several bright stars can be seen without binoculars or telescopes.

- These stars have been seen since ancient times.

- A good view of the Pleiades can be had in late autumn into winter.

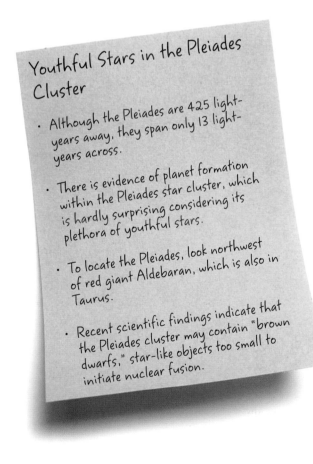

Youthful Stars in the Pleiades Cluster

- Although the Pleiades are 425 light-years away, they span only 13 light-years across.

- There is evidence of planet formation within the Pleiades star cluster, which is hardly surprising considering its plethora of youthful stars.

- To locate the Pleiades, look northwest of red giant Aldebaran, which is also in Taurus.

- Recent scientific findings indicate that the Pleiades cluster may contain "brown dwarfs," star-like objects too small to initiate nuclear fusion.

the grouping appear rather faint to our observation and why so many of them cannot be seen without a telescope.

As you could easily surmise, the Pleiades cluster, with its famous nickname of the "Seven Sisters," has seven corresponding stars: Sterope, Merope, Electra, Maia, Taygeta, Caleano, and Alcyone. The latter star is the brightest of the lot. In fact, Alcyone is estimated to shine with an intensity one thousand times brighter than that of our sun. From our unique vantage point here on Earth, distance really matters vis-à-vis locating deep-sky objects. It impacts not only what we see but also how bright the objects are when we do see them. Oh, in addition to the seven main stars in the cluster, the sisters, there are two more major stars—the sisters' parents, as it were: Atlas and Pleione.

The Pleiades cluster is an absolute must-see. When you zero in on this tightly knit group of stars, the individual stars seem sheathed in warm blue light. This is the byproduct of their powerful luminescence reflecting off the copious amounts of dust in this region of space.

Visible to the Naked Eye

- The Pleiades are simultaneously among the brightest and nearest of open star clusters.

- In the annals of amateur astronomy, it's fair to say that the Pleiades are the best known of open star clusters and have been since antiquity.

- Courtesy of its relative closeness to Earth, it's unquestionably the most striking star cluster to the unaided eye.

- Blue reflection nebulae encase many of the stars in an ethereal glow.

The Seven Sisters

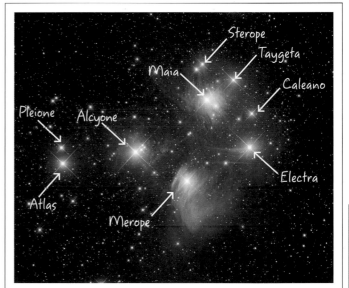

- Also known as the "Seven Sisters" from Greek mythology, the Pleiades star cluster actually flaunts nine bright stars.

- The Seven Sisters in mythology are Sterope, Taygeta, Maia, Caleano, Alcyone, Electra, and Merope.

- The two additional stars visible in the night sky are Pleione and Atlas, the mythological Seven Sisters' mom and dad.

- Alcyone, a binary star, is the brightest within the cluster. Three companion stars orbit it: Alcyone B, Alcyone C, and Alcyone D.

BEEHIVE CLUSTER

The Beehive cluster includes many aging and bright stars in a neat symmetry

The Beehive cluster, M44, is an open cluster of stars in the constellation Cancer the Crab. It is often referred to as the "Praesepe," which is the Latin term for "manger." This deep-sky object is among the closest star clusters to our solar system and thus provides stargazers with a fair peek into its luminous boundaries. In fact, among the first deep-sky objects that Galileo explored with his fledgling telescope was the Beehive cluster.

The Beehive cluster is approximately 580 light-years from Earth. It spans 10 light-years in diameter. With a heavy concentration of red giant stars and white dwarf stars, the cluster is seen as being in its latter half of evolution. That is, these

The Praesepe

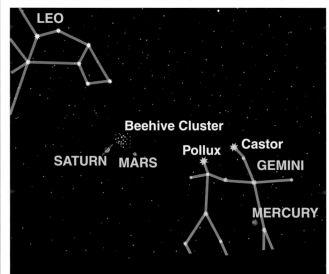

- The Beehive cluster, M44, is an open cluster of stars found in Cancer the Crab. It is frequently referred to as the "Praesepe."

- The cluster is situated in the center of an overall dim constellation of stars.

- It is roughly midway between Regulus in zodiac neighbor Leo and Pollus in zodiac neighbor Gemini.

- Although the Beehive cluster can be seen with the naked eye, binoculars will supply you with a more defined look.

Telescopes and Open Clusters

- The Beehive cluster has been known since antiquity. It accommodates a greater star population than do most naked-eye open clusters. With a telescope, multiple hundreds of stars come into view.

- Massive young stars congregate near the cluster's center.

- Although it is visible to the naked eye under optimal conditions, the cluster's stars are not especially bright. Light pollution will obscure the view.

- The Beehive cluster is approximately 580 light-years away.

types of stars are old in stellar terms, dying, as a matter of fact, and losing their capacities to generate nuclear fusion. Red giant stars tend to grow brighter as they shed their outer layers and swell in size. White dwarf stars are what the red giants ultimately become when they deplete their last vestiges of hydrogen, leaving behind only piping hot cores of helium gas.

The Beehive cluster includes both immense and intense stars near its center. Fainter, less-substantial stars inhabit the cluster's so-called halo or corona. The star 42 Cancri is in the Beehive cluster.

NGC 2158

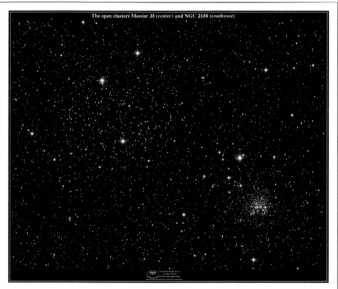

The open clusters Messier 35 (center) and NGC 2158 (southwest)

Invisible to the Naked Eye

- Like the Beehive cluster, NGC 2158 is an open cluster of stars.

- Whereas the Beehive cluster can be found in Cancer, NGC 2158 calls home the neighboring zodiac constellation of Gemini.

- NGC 2158 is in close proximity to a more popular open cluster in Gemini: M35.

- NGC 2158 is an elderly cluster of stars believed to be more than 1 billion years old.

- To the naked eye, open star clusters usually appear as mostly dark areas of the night sky. Binoculars will add more stars to the picture and telescopes even more.

- Telescopic views of the Beehive cluster and other open clusters expose hundreds of stars.

- Ancient Romans called the Beehive cluster the "Praesepe," the Latin word for "manger," because they thought the star pattern resembled two donkeys feasting on hay.

- The "donkeys" were the two stars Asellus Borealis and Asellus Australis.

WHIRLPOOL GALAXY

The Whirlpool supplies astronomers with a fascinating picture of a spiral galaxy in action

The Whirlpool galaxy, M51, is a spiral galaxy approximately 23 million light-years from Earth. It is located in the constellation Canes Venatici. With a good pair of binoculars, it is readily spotted in the sky. As you might imagine from its name, the galaxy showcases a very explicit spiral composition. That is, it resembles an interstellar whirlpool with its two prominent

arms spinning around its galactic center. These arms are actually elongated corridors of gas and stars commingling with dust. However, what you will observe through your binocular or telescopic lenses will not supply you with the depth that long astrophotographic exposures reveal. These pictures expose a fair share of the faraway galaxy's definition.

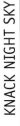

Famous Galaxy

- The Whirlpool galaxy, M51, is unquestionably one of the most scrutinized and famous of all galaxies.

- Its two prominent arms, which majestically sweep around its bright yellow core, are loaded with hot blue stars.

- Charles Messier discovered the galaxy and included it in his catalog of deep-sky objects. He branded it "a very faint nebula without stars."

- The optimal time to hunt down the Whirlpool galaxy is in the spring.

Whirlpool's Stellar Companion

- The Whirlpool galaxy has a faithful companion at its side: NGC 5195. This much smaller galaxy is slowly but surely making its way behind the Whirlpool.

- The nearness of this tiny galactic passerby nonetheless impacts the sprawling Whirlpool galaxy, generat-

ing evident ripple effects within it.

- Clouds of dust and gas are kicked up in what resemble stormy skies.

- Pinkish areas in the Whirlpool galaxy reveal the locations of star formation.

It is believed that the Whirlpool's galaxy's tightly wound organization is the consequence of its close proximity with a companion galaxy (NGC 5195), which looks like it is tugging on the Whirlpool's outer edges. It is, in fact, behind the Whirlpool galaxy and not in any danger of an intergalactic collision. Nonetheless, this celestial neighbor's gravitational pull is fueling star formation within the Whirlpool itself. Bright clusters of stars, which indicate stellar youth, are omnipresent on the galaxy's fringes. Radiant blue-colored starlight is highly visible—a telltale indicator of hot young stars at play.

Astronomical inspections of this intriguing galaxy reveal, too, what looks like dust "spurs," as they are called, branching out from its two main spiral arms. This had engendered speculation that the Whirlpool galaxy may, in fact, have multiple spiral arms at work. In the galaxy's center, a dust disk is discernible, leading to more scientific conjecture that it is supplying sustenance to a lurking and mammoth black hole.

Impressive Spiral Arms

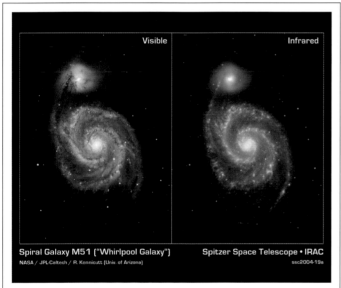

Spiral Galaxy M51 ("Whirlpool Galaxy")
NASA / JPL-Caltech / R. Kennicutt (Univ. of Arizona)
Spitzer Space Telescope • IRAC
ssc2004-19a

- The spiral arms of the Whirlpool galaxy are unmistakable. The high concentration of dust in these arms is apparent.

- The gaps between the arms are also visible.

- The Whirlpool galaxy can be found in the constel-lation Canes Venatici. The naked eye won't cut it with this one, but a telescope will give you a fair peek.

- To find the Whirlpool galaxy, first locate the tail star of Ursa Major the Great Bear. Trace a path from this star, known as "Benetnash," to the southwest.

X Marks the Spot

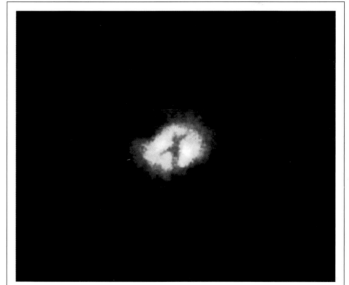

- The Hubble space telescope's image of the Whirlpool galaxy's core exposes what scientists believe is a black hole.

- X marks the spot in this picture. It is prominently seen within the galaxy's nucleus—a bona fide point of no return.

- Scientists further speculate that the black hole could be as massive as one million suns.

- The Whirlpool galaxy is approximately 100,000 light-years in diameter and 31 million light-years from Earth.

GREAT GLOBULAR CLUSTER

This cluster is a mass of stars bound together by compelling gravitational forces

The Great Globular cluster, M13, can be found in the constellation Hercules. It is sometimes referred to as the "Great Globular cluster in Hercules" or "Hercules Globular cluster." The cluster is 20,000 light-years from Earth and estimated at 145 light-years in diameter. As a rule, globular clusters consist of spherical congregation of aging stars with a common ancestry, if you will.

The Great Globular cluster, like fellow globular clusters, orbits a dominant galactic center. In this case, our very own Milky Way's.

Globular clusters reside in their parent galaxy's halo or galactic bulge. The stars, which total hundreds of thousands, in the Great Globular cluster are securely bound together by the Milky Way's incredible gravitational pluck. This is what

A Favorite Deep-Sky Object

- The Great Globular cluster in Hercules, M13, is among the most sought-out deep-sky objects in the summer night sky.

- Under dark and otherwise optimal conditions, the cluster can be spotted with the naked eye. However, binoculars and telescopes are recommended for viewing it.

- The Great Globular cluster is located in the constellation Hercules, spans about 145 light-years, and is 25,000 light-years away.

- The cluster is believed to house hundreds of thousands of stars.

Dense Grouping of Stars

- The Hubble space telescope's close-up view of the Great Globular cluster in Hercules is revealing. The incredible density of the star cluster is obvious.

- The spherical shape found in globular clusters is also on display here. This is the consequence of our galaxy's persuasive gravitational muscle.

- Globular clusters exist in large numbers in the Milky Way's halo.

- The Great Globular cluster is just one of more than two hundred such clusters observed.

sets the Great Globular cluster apart from conventional star clusters, which are less at the beck and call of the forces of gravity and more loosely dispersed in outer space.

In the Great Globular cluster's case, we are talking about a close-knit interstellar community of stars. It is presumed that stars near a galaxy's core are close to five hundred times more concentrated than stars existing elsewhere. The Milky Way at large is presumed to have at least 250 globular clusters within its boundaries and likely many more than that. Other galaxies' globular clusters run in the thousands and even millions.

Home to Stars of All Ages

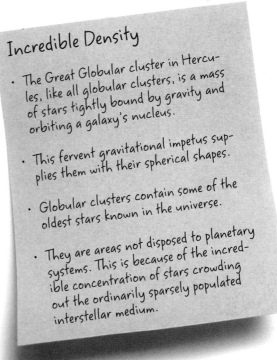

Incredible Density

- The Great Globular cluster in Hercules, like all globular clusters, is a mass of stars tightly bound by gravity and orbiting a galaxy's nucleus.

- This fervent gravitational impetus supplies them with their spherical shapes.

- Globular clusters contain some of the oldest stars known in the universe.

- They are areas not disposed to planetary systems. This is because of the incredible concentration of stars crowding out the ordinarily sparsely populated interstellar medium.

- The blue and white stars evident in the Great Globular cluster of Hercules are the children in this celestial house.

- Globular clusters, including M13, simultaneously accommodate some of the oldest stars in the universe. These stars, radiating bright red and orange, are red giants.

- Courtesy of the ultracrowding in the cluster, inhabitants have been known to butt heads on occasion and form new stars, which are called "blue stragglers."

- Blue stragglers are hot and bright stars.

DEEP SKY: SPRING-SUMMER

RING NEBULA

This planetary nebula is a multihued derivative of a dying star not unlike our sun

The Ring nebula, M57, is indisputably among the most intriguing deep-sky objects. With a pair of binoculars or a wide-lens telescope, you can see, in living color, this remarkably multihued celestial body. The Ring nebula is located in the northern constellation Lyra the Harp. It is rather easy to locate because this sizeable gaseous space entity is due

south—a hop, skip, and a jump—from Lyra's brightest and well-known star, Vega.

The Ring nebula is classified a planetary nebula. And this designation does not mean that it is in any way, shape, or form related to, or derivative of, a planet. In fact, the label is strictly an appearance thing. The Ring nebula, like all other

Stargazing Favorite

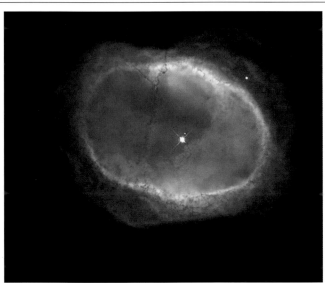

- The Ring nebula, M57, is classified a planetary nebula, although these kinds of nebulae have nothing to do with planets. They are formed when medium and low mass stars, like our sun, run dry of their hydrogen fuel sources.

- The Ring nebula can be readily found. First find Vega, the brightest star in Lyra. It is not too far south from this prominent star, which is part of the Summer Triangle.

- Summer is the best time to check it out.

Shapely Nebula

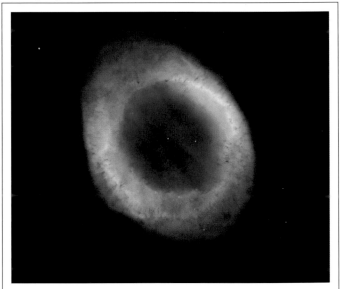

- The Ring nebula is considered a quintessential planetary nebula because of its defined shape and colorful gases.

- Responsible for the glowing nebulosity is a dying star at the nebula's center.

- The outer layers of the star have been expelled into the celestial ether.

- The gases' myriad colors tell a tale with the blue—very hot helium—close to the central star. Green indicates ionized oxygen, whereas red, at the farthest distance from the center, is ionized nitrogen.

planetary nebulae, came to pass upon the demise of a low to medium mass star. Our sun fits such a bill. That's really the long and short of it. When this particular star depleted its hydrogen fuel source, which once upon a time initiated nuclear fusion at its core, it remade itself to survive. That is, this rather ordinary star altered its structure and continued living, so to speak, by dispensing with its outer layers. Essentially, it expanded into a red giant star—very large and very bright—but nonetheless weakening along the way.

Effectively, during this star's stellar death rattle, a gaseous outer shell formed. One that is clearly visible as a spherical nebula in space, which is lit up by the ultraviolet energy still existing in the star's feverish core. The Ring nebula is, in fact, a prime example of one dying star's colorful transition from red giant star to white dwarf star. It is an amateur astronomer's catch with its multicolored hues of blue, yellow-green, and red. For the record: The blue is ionized helium; the yellow-green, doubly ionized oxygen; and the red, ionized nitrogen.

Ring Nebula in Infrared

- The Ring nebula's gaseous veil is clearly evident in this infrared image.

- The nebula's clouds of gas and dust are also visible well beyond its ring-shaped innermost regions.

- It is difficult to assess the distance of the Ring nebula, but 2,000 to 4,000 light-years away is a consensus estimate. It is only 1 light-year in diameter.

- After the planetary nebula dissipates, which it will eventually, it will be a faint white dwarf star.

Close Encounters

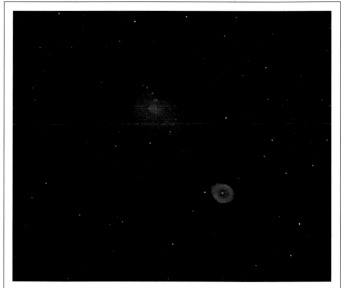

- In this compelling celestial shot, a comet's tail appears to glance the Ring nebula.

- The Ring nebula is not spherical in shape but rather cylindrical.

- A small telescope will supply you with a nice view of the Ring nebula. You will be able to make out its unusual contours and colors but not its central star.

- Even powerful telescopic inspections of the nebula have difficult times finding the star behind it all.

DUMBBELL NEBULA

The first planetary nebula discovered, the Dumbbell, is a favorite target of stargazers

Just like the Ring nebula, the Dumbbell nebula, M27, is a planetary nebula. However, it's a decidedly different visual for stargazers, with distinctive characteristics on display. The Dumbbell nebula can be found in the constellation Vulpecula. It is approximately 1,360 light-years from Earth and 975 light-years in diameter. It also has the distinction of being the first planetary nebula discovered. Charles Messier first laid eyes on it in 1764. He called it a "nebula without a star," although a core white dwarf star exists.

Although the Dumbbell nebula is neither the biggest nor brightest deep-sky object of its kind, it is one of the most accessible to amateur astronomers with binoculars or a

First Planetary Nebula Discovered

- The first planetary nebula discovered is called the "Dumbbell nebula," M27. It is located in the constellation Vulpecula.

- Among amateur astronomers, the Dumbbell nebula is considered a must-see deep-sky object.

- It is estimated to be approximately 1,200 light-years away and to span 2.5 light-years in diameter. However, let's just say that measuring nebulosity is not an exact science.

- The Dumbbell nebula is considered an excellent telescope target.

Brightest in the Night Sky

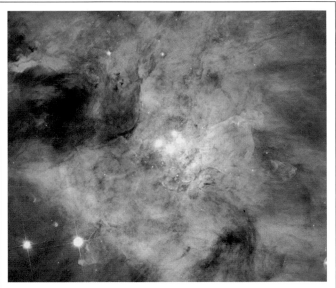

- The Dumbbell nebula is among the brightest planetary nebulae visible in the night sky.

- The central star, which is the wind beneath the nebula's wings, is faint in the image.

- Nonetheless, it is this dying star's ultraviolet light, emanating from its intensely hot core, that is heating and exciting the gases of its shedding layers.

- The expelled shells of the star fluoresce and supply the nebula with its multicolored appearance.

small telescope. With either of these enhancing devices, the Dumbbell nebula can be seen in vivid detail with a pale green tone lacing its outer edges, rounded contours, and an unusually knotty interior.

It is, in fact, ultraviolet light emanating from the nebula's core that triggers, when interacting with oxygen, this ethereal green glow. The red shades are the byproduct of its commingling with hydrogen. The Dumbbell nebula supplies an elegant celestial image of a planetary nebula and how it came to be.

ZOOM

Planetary nebulae originate rather quietly from the remains of low and medium mass stars. They are decidedly different interstellar occurrences than supernova events. Supernovae are epic explosions of weighty, massive stars. Planetary nebulae's formations are positively placid by comparison—no super explosions or big bangs.

Pronounced Halo

- A pale but substantial halo can be seen surrounding the Dumbbell nebula.

- The Dumbbell nebula received its nickname because of its elongated contours, which evidently resembled dumbbell weights to someone.

- Its middle region reveals a pattern of knotting, which is common with planetary nebulae.

- Although the white dwarf star responsible for the Dumbbell nebula appears dim, its radius is the largest of its kind. It is one of the brighter white dwarf stars.

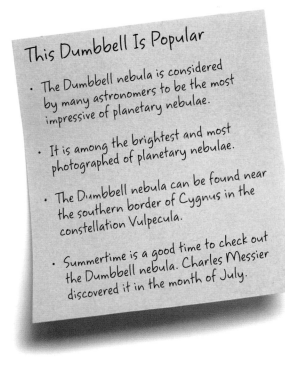

This Dumbbell Is Popular

- The Dumbbell nebula is considered by many astronomers to be the most impressive of planetary nebulae.

- It is among the brightest and most photographed of planetary nebulae.

- The Dumbbell nebula can be found near the southern border of Cygnus in the constellation Vulpecula.

- Summertime is a good time to check out the Dumbbell nebula. Charles Messier discovered it in the month of July.

LAGOON NEBULA

The Lagoon nebula is a breeding ground for many of the stars of tomorrow

The Lagoon nebula, M8, is a vast interstellar cloud of dust and gas. It is located in the zodiac constellation of Sagittarius and is some 4,100 light-years from Earth. It is not among the subclassification "planetary nebulae," and so it is a different space animal altogether from the Ring nebula and Dumbbell nebula. The Lagoon nebula is instead known as an "emission nebula," which is an area in space highly disposed for star formation.

The Lagoon nebula can be adequately seen with binoculars and a basic telescope. Observers will note that this deep-sky object sports a very defined oval form and appears as an opaque patch of clouds in the dark night sky. Particularly

Emission Nebula

- The Lagoon nebula, M8, can be found in Sagittarius. It is considered a huge interstellar cloud and dubbed an "emission nebula."

- It appears gray when observed through binoculars and a telescope but red and pinkish in time-exposure color photographs.

- The Lagoon nebula is barely visible with the naked eye and only under optimal conditions.

- Binocular observation brings the Lagoon nebula into focus. You can make out its oval contours and a highly defined epicenter.

Fertile Star-Forming Turf

- The Lagoon nebula's gaseous clouds surround young stars. This region of the night sky is among the brightest and most fertile star-forming locales.

- The nebula has already fathered its own cluster of hatchling stars: NGC 6530.

- A darkened pathway runs through the nebula. This visibly winding lane is where the name *lagoon* comes from.

- It is situated north of the Sagittarius star cloud, or just west of the top star in the constellation's well-known Teapot asterism.

conspicuous is the Lagoon nebula's well-lit center. It is here that scientists believe life is being breathed into countless stars of tomorrow.

In this ultrabright region of the Lagoon nebula is a secondary nebula, which has been dubbed the "Hourglass nebula" because of its shapely figure. The Hourglass nebula is visible in the area of the Lagoon nebula that is ground zero for the aforementioned star breeding. Indeed, scientists consider this epicenter of the Lagoon nebula a veritable space laboratory, a microcosm of the perpetual dynamism of the universe.

Twisters

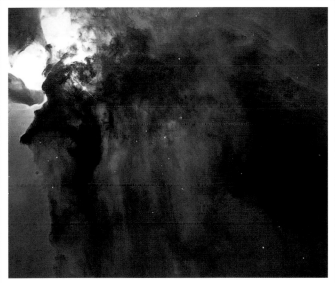

- The Hubble space telescope has captured a striking snapshot of the Lagoon nebula, which reveals a couple of conspicuous "twisters" in its center region.

- The twisters span a distance of .5 light-year.

- They appear to be funneling away from the nebula's central star.

- The presence of what are called "dark globules" in the Lagoon nebula intrigues scientists. Behind the formation of stars, these globules are contracting protostellar clouds.

Nebula in a Nebula

- The Hourglass nebula is seen within the brightest region of the Lagoon nebula.

- This is an area of the Lagoon nebula that appears to be exceedingly active in star formation.

- The radiance in the Hourglass nebula environs is the byproduct of extremely hot and extremely young stars.

- The Hourglass nebula itself is lit up courtesy of a bright star, Herschel 36. Herschel was hatched in the Lagoon nebula's stellar nursery.

WHAT ARE COMETS?

Comets are a mixture of frozen gases, dust, and rocky materials that forever fascinates us

Comets making their way through outer space and revealing themselves to us only sparingly have developed a celestial aura that is unmatched. They never cease to warrant our attention. Comets aren't stars, planets, moons, or nebulae, which are always where we expect to find them at certain moments during the night and particular times of the year.

Indeed, even people with minimal interest in astronomy take note of comet appearances and try to catch a glimpse.

Comets are distinct celestial objects with—in cosmological terms—short life spans. They are a unique blend of frozen gases, dust, and solid materials. When a comet loses its volatility over time—via melting and evaporating gases—it

KNACK NIGHT SKY

Comet West

- Comet West was christened the "Great Comet of 1976."

- Its multiple tails were resplendent against a backdrop of stars. The comet's plasma tail highlights the presence of hot gases, whereas its more expansive multiple tails illuminate dust particles.

- Comet West's prior appearance in the night sky was 550,000 B.C. If you weren't around to see it in 1976, you'll have to content yourself with pictures.

- Comet West's orbital period is presumed to be more than 500,000 years.

Comet McNaught

- Comet McNaught received the title "Great Comet of 2007."

- It is deemed a long-term comet, which means that its orbit around the sun is two hundred years or more. McNaught's actual orbiting time is nonetheless unknown.

- Around perihelion—its point closest to the sun—McNaught could be seen during the daylight hours all across the world.

- Extraordinary solar heat enhanced Comet McNaught's brightness for one brief shining moment.

morphs into just another piece of rock aimlessly meandering through space. Scientists believe that many asteroids are, in fact, former comet nuclei.

Comets travel in sweeping elliptical orbits around the sun, which takes them both very deep into our solar system and very near the sun. Their contours are highly irregular—no two are the same. A comet's head and atmosphere are typically an expanding and fuzzy cloud of widely scattered materials, referred to as its "coma." The engine of a comet is at the center of the coma: a small but nonetheless radiant and solid

nucleus. An individual comet's coma will typically swell in size and brightness as it nears the heat and strong solar wind of the sun.

In essence, the coma and nucleus constitute the comet's body. But often the coma leaves behind visible tracks of gas and dust in its wake as it journeys near the sun. This residue is known as the comet's "tail" and can look very impressive when illuminated by the sun and observed from Earth. Some comets sport multiple tails. Their tails are the stuff of legend.

Comet Holmes

- Since its discovery by an amateur astronomer in 1892, Comet Holmes had long been regarded as a dim, pedestrian celestial object.

- Its brightness exploded by a factor of one-half million or even more during its appearance in 2007.

- This unanticipated flare-up was the most dramatic ever observed in a comet.

- For a brief period, as its coma swelled, Comet Holmes became the biggest thing in our solar system.

Nucleus and Coma

- The circular area around the comet's nucleus is called its "coma." This sphere consists of myriad gases.

- The nucleus and coma form the noticeable head of the comet with its tails trailing behind as it approaches the sun.

- Upon exiting the sun's turf, the tail will lead the comet.

- The solar wind determines the comet tails' directions vis-à-vis its nucleus and coma. The tails will fade fast as the comet leaves the sun's hinterland.

COMET OBSERVATION

Comets make brief and rare appearances as they complete long orbits around the sun

Since 1759, the renowned comet, Halley, has made only three appearances: in 1835, 1910, and 1986. Scientists estimate that these intervals between appearances—seventy-five and seventy-six years, respectively—are about right and predict its next showing will be around 2062. But the reality is that each and every comet is a unique celestial body. They take widely varying amounts of time to orbit the sun.

The astronomical community classifies comets in two categories: short-term comets and long-term comets. Short-term comets require less than two hundred years to complete one trip around the sun. Long-term comets need more time than that. Halley's Comet is a short-term comet.

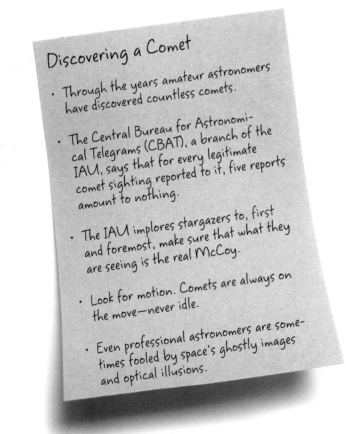

Discovering a Comet

• Through the years amateur astronomers have discovered countless comets.

• The Central Bureau for Astronomical Telegrams (CBAT), a branch of the IAU, says that for every legitimate comet sighting reported to it, five reports amount to nothing.

• The IAU implores stargazers to, first and foremost, make sure that what they are seeing is the real McCoy.

• Look for motion. Comets are always on the move—never idle.

• Even professional astronomers are sometimes fooled by space's ghostly images and optical illusions.

Two Tails

• A comet in the distant sky distinguishes itself with its blue plasma tail and even longer and broader dust tail.

• Its melting nucleus and evaporating volatiles have generated a considerable gaseous coma around it.

• The gases in the coma soak up ultraviolet radiation. This supplies the comet with an alluring fluorescent glow.

• Flourescence in outer space furnishes a more captivating radiance than mere reflected light.

There are untold numbers of comets in outer space, and the vast majority of them are too tiny or dim to be seen without the aid of a telescope. The comets that reveal themselves—sometimes in grand fashion—to the naked eye are passing near the sun. What we see is their gaseous and dust-laden nuclei and atmospheres reflecting the potent sunlight back to Earth. And, too, as these comets' gases interact with the sun's energy and stiff solar wind, a fluorescent glow is often the consequence. This all makes for an impressive picture in the night sky and even sometimes in the daytime sky when the comets are especially luminescent. Comets' reappearances on Earth's stage usually last several weeks.

In 1997 the Comet Hale-Bopp came rather close to Earth—a mere 122 million miles (197 million kilometers) away. This skirting of Earth, as it were, is not really unusual for comets. But Hale-Bopp's considerable nucleus, approximately 18 to 25 miles (30 to 40 kilometers) in diameter, spewed an inordinate amount of gas and dust in its wake. And, as the sun illuminated the extensive debris, Hale-Bopp appeared very bright and highly visible to the unaided eye.

Comet Bradfield

- Cutting quite a swath across the night sky, Comet Bradfield's elongated tail made for quite a visual.

- Longtime comet hunter William Bradfield's diligence has certainly paid off. Bradfield has discovered multiple comets through the years.

- Individuals specifically on the prowl for comets are behind most of their discoveries. That is, men and women systematically combing selected snippets of the night sky with their telescopes.

- Binoculars are ideal for zeroing in on comets.

Comet Kohoutek

- Czech astronomer Lubos Kohoutek discovered Comet Kohoutek in 1973. It was visible into 1974.

- As has so often been the case in recent times, comets' appearances are frequently overhyped by the media.

- Comet Kohoutek was heralded as the "Comet of the Century" and didn't live up to its billing.

- Some cynical folks dubbed it "Comet Watergate," equating its lackluster showing with the political scandal wracking the nation at the same time.

COMET HYAKUTAKE
An amateur astronomer discovered the Great Comet of 1996

Comet Hyakutake was justly christened the "Great Comet of 1996." What made this comet's appearance so spectacular in March was its recent discovery. That is, Hyakutake wasn't an anticipated comet like Hale-Bopp soon after in 1997 or Halley's Comet in 1985. No, Hyakutake was first observed and recorded on January 31, less than two months before it made news. And on top of everything else, an amateur astronomer saw it first. Employing a powerful pair of binoculars, Yuji Hyakutake of Japan spotted the comet before anybody else and was rewarded for his good fortune when the comet took his name.

Comet Hyakutake was witnessed around the world. Its approach vis-à-vis Earth was closer than that of any comet in almost two hundred years, and the brightest comet to come our way since Comet West in 1976.

It was soon determined that Hyakutake was a long-term

Comet Hyakutake Hugs the Horizon

- Comet Hyakutake ambles dramatically close to the horizon in the early evening hours.

- Hyakutake proved to be one of the closest comet visitations in two centuries. This celestial box seat supplied astronomers with an intimate peek at a comet in action.

- Surprised, scientists found the comet emitting X-rays, the first such discovery.

- An amateur astronomer discovered Hyakutake in the winter of 1996.

Super-Bright Nucleus

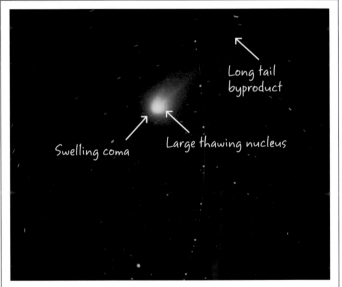

Long tail byproduct

Swelling coma

Large thawing nucleus

- Comet Hyakutake's sizeable and bright nucleus distinguished it in the fraternity of comets.

- As Hyakutake orbited close to the sun, the comet's extensive tails of gas and dust were also impressive.

- By chance, NASA's *Ulysses* spacecraft crossed the path of Comet Hyakutake and discovered that its vaunted tail was even bigger than originally surmised.

- The probe determined that the tail extended more than 300 million miles (500 million kilometers)—the longest comet's tail ever seen.

comet, meaning that its orbit around the sun takes more than two hundred years to complete. In Hyakutake's case, it was estimated at seventeen thousand years since it last graced our nighttime skies. So, people of the Stone Age likely got a fair look at Hyakutake. What they thought when they saw this celestial concoction streaking across the sky is anybody's guess. But, here's the real kicker: This comet is not expected to show itself again for 100,000 years. Courtesy of increased gravitational influences of the giant planets in our solar system, which are slowing Hyakutake down considerably, it's going to take 100,000 years now to orbit the sun. In other words, this comet has got a long slog to the outer rim of our solar system and back before we will see it again.

Elongated Blue Ion Tail

- Comet Hyakutake's blue ion tail is a sight to see in the night skies.

- Hyakutake bowed off the celestial stage in May 1996 and is not expected to return to Earth's view for another 100,000 years.

- Until Hyakutake came along, the Great Comet of 1843 had the longest comet tail on record.

- In fact, Comet Hyakutake smashed the all-time tail record with miles to spare, practically doubling the span of the Great Comet of 1843.

Comet Hyakutake Close-Up

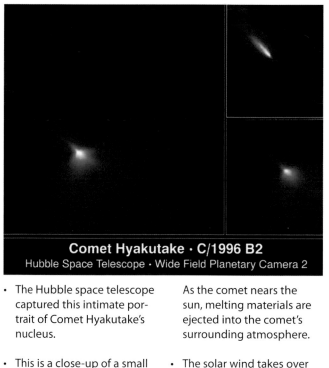

Comet Hyakutake · C/1996 B2
Hubble Space Telescope · Wide Field Planetary Camera 2

- The Hubble space telescope captured this intimate portrait of Comet Hyakutake's nucleus.

- This is a close-up of a small area in proximity to the comet's icy nucleus.

- This icy region of Hyakutake is responsible for the tail.

- As the comet nears the sun, melting materials are ejected into the comet's surrounding atmosphere.

- The solar wind takes over from there and blows gas and dust particles away from the comet's coma.

COMET HALE-BOPP
The bright Comet Hale-Bopp is the most widely seen comet of all time

Comet Hale-Bopp is undeniably the most celebrated comet of recent times. When it appeared for all to see in 1997, more people caught a glimpse of Hale-Bopp than any other comet before or since. Indeed, Hale-Bopp was visible in every quarter of the planet. In fact, its debut lasted a record-smashing eighteen months, which shattered the previous record of

nine months held by the Great Comet of 1811.

Hale-Bopp richly earned its title as the "Great Comet of 1997." Aside from being observable in the sky for an extensive length of time, this comet was remarkably bright. So bright that it was visible in big cities cloaked in light pollution, which is frequently the bane of amateur astronomers.

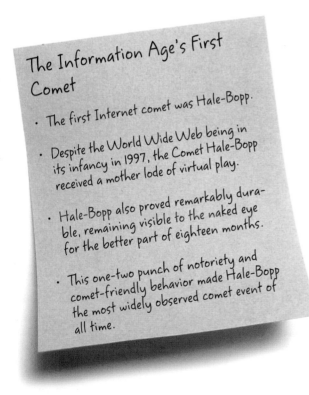

The Information Age's First Comet

- The first Internet comet was Hale-Bopp.

- Despite the World Wide Web being in its infancy in 1997, the Comet Hale-Bopp received a mother lode of virtual play.

- Hale-Bopp also proved remarkably durable, remaining visible to the naked eye for the better part of eighteen months.

- This one-two punch of notoriety and comet-friendly behavior made Hale-Bopp the most widely observed comet event of all time.

Trail of Two Tails

- Comet Hale-Bopp's two distinct tails—one blue and one white—are on full display.

- Stargazers all across the world caught a glimpse of this impressive comet, which was both highly visible and around for a long time.

- Hale-Bopp was justifiably christened the "Great Comet of 1997."

- Whereas an average comet's nucleus is 3 miles (5 kilometers), astronomers estimated Hale-Bopp's nucleus to be in the neighborhood of 18 to 25 miles (30 to 40 kilometers).

Its ultrabrightness was due to its considerable nucleus that spanned 18 to 25 miles (30 to 40 kilometers) in diameter. The average comet's core is only 3 miles or even smaller.

Like Hyakutake, which showed itself in the night skies the year before, Hale-Bopp is a recently discovered comet. In 1995 two independent observers of the celestial beyond, Alan Hale and Thomas Bopp, spotted the comet before anybody else. Its subsequent advance into the inner regions of our solar system was thereafter widely anticipated but was briefly upstaged by the big comet surprise, Hyakutake, which seemed to materialize from nowhere. Despite its competition, Hale-Bopp didn't disappoint, with luminosity one thousand times stronger than Halley's Comet at a comparable distance. Scientists also discovered that Hale-Bopp sported multiple tails, including one consisting of sodium. Those of you who didn't get a chance to see Hale-Bopp with your own two eyes will have to be content with the photographs taken by others. Hale-Bopp's expected return to Earth's view will be in the year 4397. Yes, Hale-Bopp is a long-term comet, too.

Comet Hale-Bopp Closing in on the Sun

- Comet Hale-Bopp's expanding tails are in full bloom as it ventures closer and closer to the sun.

- Hale-Bopp's white dust tail is quite pronounced, and its blue ion tail is equally impressive.

- Although Hale-Bopp didn't get any nearer to the sun than Earth does, the solar wind's impact is more dramatically felt.

- In the neighborhood of the sun, the comet's massive and frozen core excretes melting gases and dust particles into space.

Super-Sized Nucleus

- Scientists suspect that Hale-Bopp's nucleus measured three times the size of the comet or asteroid that crashed into Earth and decimated the dinosaur population.

- This celestial body's nucleus likely measured 6 to 9 miles (10 to 14 kilometers).

- Hale-Bopp was a considerable comet that came very near Earth.

- Hale-Bopp frequently overwhelmed the luminescence of nearby stars during its appearance. At its brightest, it could even be seen from big cities.

COMET HALLEY

Halley's Comet is the only short-term comet that is visible to the naked eye

Comet Halley—or "Halley's Comet," as it's often called—is without question the most famous short-term comet in history. In fact, it's the only short-term comet—that is, a comet that periodically appears in Earth's night skies in under a two hundred-year time frame—that is discernible to the naked eye. Contrarily, many long-term comets, including Hale-Bopp and Hyakutake, are easily spotted without binoculars and a telescope. But not so with most short-term comets, which tend to be less bright.

For what it's worth, Halley's Comet is the only known comet that can be observed more than once in a human being's lifetime. It has a pronounced elliptical orbit around the sun. Scientists surmise that it has been doing its thing for a

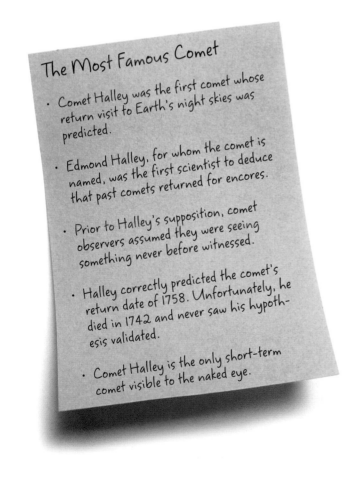

The Most Famous Comet

- Comet Halley was the first comet whose return visit to Earth's night skies was predicted.

- Edmond Halley, for whom the comet is named, was the first scientist to deduce that past comets returned for encores.

- Prior to Halley's supposition, comet observers assumed they were seeing something never before witnessed.

- Halley correctly predicted the comet's return date of 1758. Unfortunately, he died in 1742 and never saw his hypothesis validated.

- Comet Halley is the only short-term comet visible to the naked eye.

Dirty Snowball

- Halley's Comet has often been referred to as a "dirty snowball," but "dirty mud ball" is more apropos.

- Close scientific evaluation of the comet during its appearance in 1986 revealed that Halley is actually black, although we see a shimmering white.

- In its journey of seventy-five to seventy-six years around the sun, Halley's Comet ventures between the orbits of Mercury and Venus and eventually beyond Neptune's orbit.

- The comet's appearance has been recorded as far back as 240 B.C.

minimum of sixteen thousand years, and perhaps as much as 200,000 years. As far back as 240 B.C., ancient astronomers had observed this peculiar space body and recorded its movements. But it took English astronomer Edmond Halley to put it on the map, so to speak, when he identified the comet's periodic disposition in 1705.

Halley surmised that this comet returns to the inner regions of our solar system in a predictable period of time. He calculated that the comet appears in Earth's night skies every seventy-six years or close to it and that it would do so again.

Sightseeing Halley's Comet

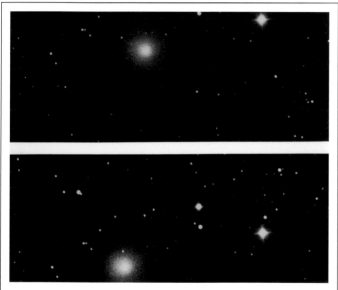

- Halley's Comet passes a globular cluster of stars on its journey around the sun.

- Halley's coma and tails are considerable in size, which is why this short-term comet is so visible to stargazers. However, its nucleus is significantly smaller than Hale-Bopp's.

- Halley's nucleus is only 9 miles (15 kilometers) long, which is less than half the size of Hale-Bopp's impressive core.

- Halley's Comet's orbit is almost perfectly elliptical.

What Makes Halley Go

- As Halley's Comet nears the sun, parts of its icy nucleus are vaporized by the sun's heating.

- The thawing ice expels gases and dust particles into space. This discharge forms the comet's head and body, otherwise known as its "coma."

- Solar wind simultaneously heats up the comet's gases and blows its myriad emissions away from the sun.

- The charged particles glow, and an ionized tail is visible along with a more far-reaching tail of dust.

COMET TYPES

Based on their orbiting habits, comets are classified as either short-term or long-term

The long and short of comets concerns their orbiting behavior. That is, how long it takes them to complete one full revolution around the sun. This orbiting time frame enables scientists to determine when an individual comet visible to us on Earth will next appear. The return engagement could be in several decades or more than one hundred thousand years.

Foremost, comets are classified as either short-term or long-term. Short-term comets take two hundred years or less time to orbit the sun; long-term comets take longer than that. There are stellar reasons for these dramatic differences in comet habits.

Short-term comets are born and bred in an area of space

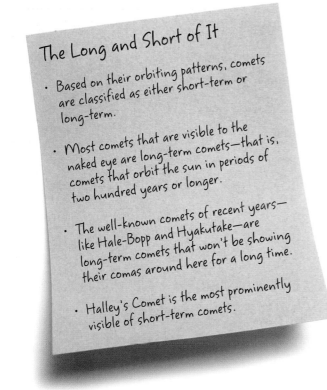

The Long and Short of It

- Based on their orbiting patterns, comets are classified as either short-term or long-term.

- Most comets that are visible to the naked eye are long-term comets—that is, comets that orbit the sun in periods of two hundred years or longer.

- The well-known comets of recent years—like Hale-Bopp and Hyakutake—are long-term comets that won't be showing their comas around here for a long time.

- Halley's Comet is the most prominently visible of short-term comets.

Comet Encke

- Comet Encke has the distinction of being the shortest of short-term comets.

- It orbits the sun every three years.

- Comet Encke's orbit inspires the annual meteor showers known as the "Taurids."

- Like most short-term comets, Encke is not easily found in the night sky. In fact, it cannot be seen with the naked eye. Binoculars are needed to spy this regular visitor.

called the "Kuiper Belt," which is situated on the outer limits of our solar system beyond the orbit of Neptune. It is considered a hotbed for short-term comets, as well as other space objects like meteoroids and asteroids. Essentially, this region of our solar system is much less dense than its inner quarters and didn't encourage planet formation. Thus, only smaller space bodies like short-term comets that orbit the sun—not sizeable planets— formed there.

Conversely, long-term comets come from a markedly different space environment known as the "Oort Cloud," which is believed to be an enormous spherical cloud surrounding our solar system. It's the presumed place in space where the sun's gravitational forces cease to be a factor. Scientists estimate this cloud at 18 trillion miles (30 trillion kilometers) away from the sun. You can thus appreciate why this area is a natural breeding ground for icy bodies like comets. In fact, this faraway sphere is the presumed home of billions of comets. When one of these bodies enters our solar system, it becomes a long-term comet and takes a meandering, and highly elliptical, journey around the sun.

Oort Cloud

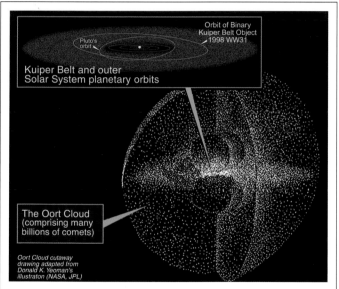

- The Oort Cloud is a hypothesized spherical cloud that surrounds the outer limits of our solar system.

- The Oort Cloud marks the celestial periphery where the sun's gravity ceases to be a player. Long-term comets enter our solar system from this distant stellar portal.

- Short-term comets orbit on the same plane as the eight planets.

Return Engagements

- Like clockwork, Halley's Comet will make a return engagement around the year 2062. You can go to the bank on it.

- But Comet Hale-Bopp is not expected to pay us a visit for another two thousand to three thousand years.

- Hale-Bopp, which glowed one thousand times brighter than Halley during its ballyhooed appearance in 1997, is a long-term comet.

- Some long-term comets take hundreds of thousands of years to complete their orbits.

WHAT ARE METEOROIDS?

Smaller than asteroids but bigger than atoms, meteoroids are solid space debris

Outer space is chock full of things careening through its immeasurable corridors. One such class of solid objects, which is bigger than atoms but conspicuously smaller than asteroids, is dubbed "meteoroids." Although this definition is rather broad, it's fair to say that the vast majority of meteoroids are minuscule—the size of a pebble or even a grain

of sand. However, some meteoroids are boulder-sized—big space rocks rattling around our solar system. But big or small, meteoroids fall into the category of space debris.

Due to their diverse shapes and sizes, meteoroids orbit the sun in a variety of manners. Estimates of their speeds in the environs of Earth's orbit are as fast as 26 miles (42 kilometers)

Space Debris

- Meteoroids enter Earth's atmosphere around the clock, burn up, and fall to the surface as dust particles.

- Meteoroids are miscellaneous space debris consisting of metals, rocks, or some combination of the two.

- The majority of meteoroids are no bigger than pebbles or sugar granules.

- Meteoroids in our solar system orbit the sun or planets. However, because of their disparate sizes, ranging from grains of sand to gargantuan boulders, meteoroids' orbits are utterly unpredictable.

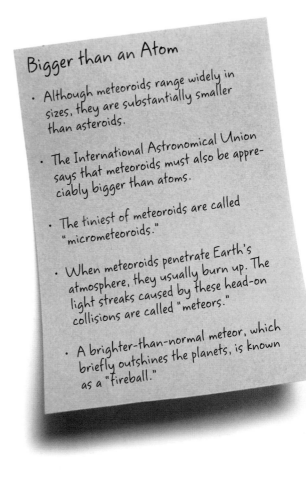

Bigger than an Atom

- Although meteoroids range widely in sizes, they are substantially smaller than asteroids.

- The International Astronomical Union says that meteoroids must also be appreciably bigger than atoms.

- The tiniest of meteoroids are called "micrometeoroids."

- When meteoroids penetrate Earth's atmosphere, they usually burn up. The light streaks caused by these head-on collisions are called "meteors."

- A brighter-than-normal meteor, which briefly outshines the planets, is known as a "fireball."

per second. And, yes, sometimes these haphazard pieces of debris, metals, and rock enter Earth's orbit. When a meteoroid physically comes into contact with our atmosphere, it officially assumes a new name: *meteor*. Colloquially, meteors are often referred to as "falling stars" or "shooting stars" and can be regularly observed in the night skies.

Most often, a meteor's touching of Earth's atmosphere signals its demise—its utter disintegration in a speedy sizzle, which leaves behind streaks of light in the night sky. In other words, meteors typically burn up before ever reaching Earth's surface.

Their crossing over into the hot upper layers of the atmosphere generates friction, which excites atoms and molecules in the vicinity, initiating luminescence that is observable from Earth.

On occasion, though, meteors do not absolutely crumble into nothingness in the atmosphere's cauldron and do find their way through. These meteors earn yet another appellation: *meteorites*. Obviously, scientists appreciate meteorites for their research potential. These are the genuine articles—space rocks—to study. Nevertheless, you wouldn't want meteorite impact on your rooftop or in your backyard.

Busy Atmosphere

- Meteoroids burn up in Earth's upper atmosphere, which also plays host to the Aurora Borealis and Aurora Australis.

- Multiple millions of meteoroids infiltrate Earth's protective shield every day.

- What we observe as meteors whizzing across the night sky are meteoroids in the frying pan of Earth's mesosphere.

- The streaks of light that dot the night sky are typically the derivative of minuscule space debris and not colossal boulders.

Meteorite Rock

- Meteorites are meteoroids that do not get totally vaporized in Earth's atmosphere and reach the surface.

- There are three very general classifications of meteorites: stony, stony iron, and iron.

- An estimated one hundred meteorites plunge into Earth's terra firma every year. Many meteorites have been identified as moon rock.

- The largest meteorite to have touched down in the United States is known as the "Willamette iron meteorite." Tipping the scales at 15 tons, it was discovered in Willamette, Oregon, in 1902.

SHOOTING STARS

"Falling" or "shooting" stars are meteors, not stars, burning up in the atmosphere

"Falling" or "shooting" stars are the stuff of legend. When we were kids, these celestial sideshows were frequently pointed out to us in the night sky. Song lyrics have long celebrated these mysterious "objects" whizzing through the heavens. How many of us have wished upon a falling star? So, we cannot be faulted for initially believing that these were, in fact,

stars. It certainly would have been less attention-grabbing to be told as children that shooting stars were, in reality, fragments of space debris—dust, metals, and rock—burning up as they crossed over into our atmosphere.

Nevertheless, these shooting stars—or "meteors," to be on scientifically solid ground—are meteoroids set ablaze on impact

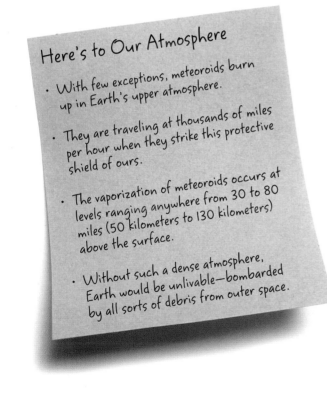

Here's to Our Atmosphere

- With few exceptions, meteoroids burn up in Earth's upper atmosphere.

- They are traveling at thousands of miles per hour when they strike this protective shield of ours.

- The vaporization of meteoroids occurs at levels ranging anywhere from 30 to 80 miles (50 kilometers to 130 kilometers) above the surface.

- Without such a dense atmosphere, Earth would be unlivable—bombarded by all sorts of debris from outer space.

Fireball

- A fireball is an exceptionally brilliant meteor.

- The American Meteor Society classifies a fireball as any meteor that shines with more luminosity than that of Venus in the morning or evening skies.

- The American Meteor Society wants you to report sightings of these fireballs, detailing their degree of brightness, color, duration, and span across the sky.

- Multiple thousands of fireballs occur daily, although most go unseen because of their location in the sky or poor timing.

with our planet's protective atmosphere. In the wake of this collision, a short-lived trajectory of light is evidence of the charring. The brightest of meteors are sometimes called "fireballs." Fireballs result from the larger, boulder-sized meteoroids—or numerous meteoroids at once—that crash into Earth's atmosphere and morph into meteors with a lot of mass to burn off.

The vast majority of meteoroids that come into contact with Earth's atmosphere are so infinitesimal that their burning away does not supply sufficient illumination to be seen from Earth.

YELLOW LIGHT

Before you go looking for meteors or meteor showers in the night sky, get the vernacular straight. Meteoroids are outerspace objects only. There are no meteoroids beyond Earth's atmosphere. When they make it into our earthly boundaries, they are meteors. And if a former space object once known as a "meteoroid," then a "meteor" makes it all the way through to Earth's surface, you've got a meteorite on your hands.

Perseids Meteor Shower

- The Perseids meteor showers occur annually in the environs of the constellation Perseus.

- Meteor showers are most often the byproduct of Earth's orbit interacting with the dusty trails of particular comets.

- Meteor showers are numerous meteor streaks of light, which appear to rain down in selected snippets of the night sky.

- The celestial location of meteor showers is known as the "radiant."

Leonid Meteor Shower

- The radiant for the Leonid meteor showers is the constellation Leo the Lion.

- Meteor showers like the Leonids have been known to generate at their peaks three hundred or more "shooting stars" per hour.

- Most meteor showers are less prolific, with twenty to thirty per hour considered a healthy showing.

- The optimal meteor shower viewing moments occur when there is a new moon, which leaves the night sky as dark as possible.

WHAT ARE ASTEROIDS?

Asteroids are solar system objects—smaller than planets—that revolve around the sun

When looking at images of asteroids what you see are rather unforgiving, eerie-looking space bodies. These malformed rock-solid objects appear devoid of any character. Asteroids are sometimes called "minor planets" or "planetoids." They are significantly smaller than planets but bigger than meteoroids. As do both planets and meteoroids, asteroids orbit the sun.

Between the orbits of Mars and Jupiter lies a region in space dubbed the "Asteroid Belt." This area bridges the last of the inner planets, Mars, with the first of the outer planets, Jupiter. The Asteroid Belt accommodates millions of asteroids within its vast frontier. However, scientists have counted no more than a few hundred with diameters surpassing 60 miles (100 kilometers).

Asteroid Ida

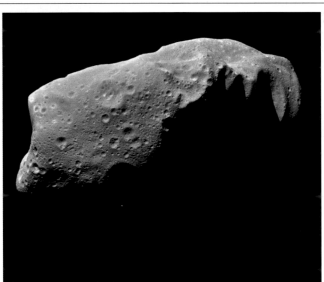

- The irregularly shaped, deeply cratered Asteroid Ida is a classic in its celestial genre.

- Ida's moon Dactyl orbits it. Dactyl's discovery marked the first asteroid moon spotted and captured on film.

- Dactyl's composition appears similar to Ida's, prompting scientific speculation that the moon and asteroid were created from the same object—that is, a larger asteroid.

- Ida and Dactyl are composed of silicate rocks.

Asteroid Belt

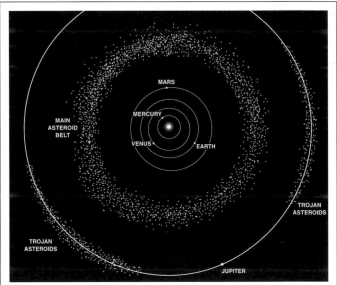

- Thousands of asteroids congregate in what is called our solar system's "Asteroid Belt." This donut-shaped area of space is located between the orbits of Mars and Jupiter.

- This area is also called the "main belt" to differentiate it from the "Kuiper Belt," which is beyond Neptune's orbit.

- Asteroids are too small to be classified as planets but are sometimes called "minor planets."

- Scientists are always keeping an eye out for asteroids that approach our planet, the so-called "near-earth objects."

They have also tallied fewer than one million with diameters at least .6 mile (1 kilometer). The rest are small by comparison but nonetheless constitute the vast majority of extant asteroids.

Asteroids are believed to be pieces of larger objects that formed planets billions of years ago. The first asteroid ever discovered is also the largest known. Its name is "Ceres." Discovered in 1801, Ceres spans 580 miles (933 kilometers).

ZOOM

Asteroids' average surface temperatures come in at approximately -100ºF (-73ºC). In the more distant regions of the Asteroid Belt, the asteroids are quite rich in carbon, which is consistent with their presumed lineage as remnants of much larger objects that formed planets. On the other hand, asteroids in the inner area of the Asteroid Belt appear more mineral laden.

Asteroid Gaspra

- The relatively smooth Asteroid Gaspra is approximately 10 miles (17 kilometers) long and 6 miles (10 kilometers) wide.

- Its gloomy gray bearing is par for the asteroid course.

- On its journey to Jupiter, the *Galileo* spacecraft passed near Gaspra. This was the closest approach of a space probe to an asteroid.

- Intriguingly, Gaspra maintains a rather widespread regolith, topography very unusual for asteroids and the subject of much scientific debate.

Near-Earth Asteroid Eros

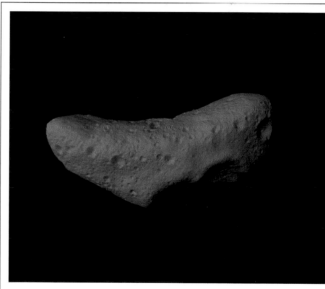

- Eros is the first observed near-earth asteroid (NEA).

- Surface gravity on Eros is highly erratic because of the asteroid's unusual shape.

- Eros is the antithesis of sphere-shaped. Astronomers have likened its contours to everything from a peanut to a boot to a potato.

- Eros is one of the few asteroids that has been paid a personal visit by a space probe. The *Near Earth Asteroid Rendezvous* (*NEAR Shoemaker*) landed on the surface of Eros in 2001.

WINTER-SPRING METEOR SHOWERS

The Quadrantids, April Lyrids, and Eta Aquarids are winter and spring meteor showers

Meteor showers are among the most spectacular of cosmic episodes. Fortunately, series of celebrated meteor showers occur every year at the same time. These meteor shower extravaganzas can be counted on because of the conventional and periodic movements of certain comets and asteroids, as well as Earth's inexorable rotation on its axis.

The Quadrantids are the year's first celebrated meteor showers as well as the last because they commence in late December. Unfortunately, this celestial event is a Northern Hemisphere show only, with the best viewing locales in the western United States. Generally, the dates of these impressive meteor showers range from December 28 through

Meteor Shower and Aurora Borealis

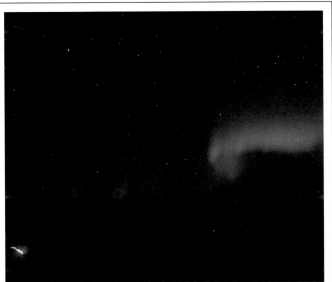

- The Quadrantid meteor showers are at once one of the most prolific and one of the least observed.

- The showers peak in early January, the heart of a Northern Hemisphere winter. The Quadrantid meteor showers are also brief. Earth's intersection with

the space debris responsible for the showers is at a perpendicular angle—that is, our planet moves rapidly through the celestial rubble.

- An asteroid by the name of "2003 EH1" is the spark that lights the Quadrantid meteor showers.

Meteor Shower Makers

- Comets streaking across the sky are simultaneously leaving trails of dusty and rocky debris in their wakes.

- When Earth's orbit intersects their beaten paths, meteor showers occur.

- Comet Thatcher is responsible for the Lyrid meteor

showers, which occur in late April.

- These meteor showers are the earliest ever recorded. Ancient Chinese records, circa 687 B.C., reported how "stars fell like rain" during a Lyrid meteor shower.

January 7. During their peak times, the Quadrantids rain down at ten to sixty meteor showers per hour. Cast your eyes to the constellation Bootes, the radiant epicenter, for a premium glimpse of these meteor showers.

The next regularly scheduled meteor showers of note occur in the springtime. Fittingly, they are known as "April Lyrids." Between April 16 and April 26, the April Lyrids can be spotted at their most prolific in the constellation Lyra. The Eta Aquarids are also noted spring meteor showers.

ZOOM

The Eta Aquarids are initiated by Earth's orbital track interacting with the dusty and rocky trail left in the wake of Halley's Comet. That is, the Earth's orbital path commingling with the famous comet's. And, interestingly, the same outer-space dynamics come to pass in the autumn, too, when two paths crossing one another inspire another annual meteor shower event, the Orionids.

Comets and Meteoroid Streams

- A comet's elongated and dusty tail is sight for stargazers' eyes. But some comets supply residual celestial seasonings, known as meteoroid streams, when they have long departed Earth's viewfinder.

- Meteor showers, like spring's Lyrids, are the byproduct of one comet's debris greeting Earth like clockwork every year at the same time.

- The radiant point for the Lyrid meteor showers is the constellation Lyra.

- This exact spot is not too far from Vega, Lyra's brightest star.

Halley's Comet and the Eta Aquarids

- Halley's Comet furnishes stargazers with a meteor shower from late April through early May.

- This annual occurrence is known as the "Eta Aquarids" and can be observed radiating from the constellation Aquarius.

- Although these meteor showers are not considered big producers, they have nevertheless been known to impress under optimal viewing circumstances.

- Meteor showers are optimally seen alongside a new moon.

SUMMER METEOR SHOWERS
The Perseids are both the most viewed meteor showers and the most spectacular

The Perseids are the best known of meteor showers. This isn't necessarily because they are the most spectacular of their kind but rather because they occur in prime time, during the dog days of summer when the maximum numbers of people are enjoying the great outdoors and checking out the night skies along the way. The Perseids reveal themselves

from mid-July through most of the month of August. Mid-August is considered the best time to witness these meteor showers at their most prolific.

Furnishing lucky stargazers with nights to remember, the Perseids have been known to peak at sixty or more meteors per hour. They are primarily visible in Northern Hemisphere

Old Reliable

- The Perseid meteor showers are annual summer events that never cease to deliver.

- The Perseids appear from mid-July through mid-August.

- The radiant can be found in the constellation Perseus.

- The Perseids put on a better show for residents of the Northern Hemisphere than Southern Hemisphere. This is because the constellation Perseus is close to the horizon and just barely shows itself to those down south.

Nature's Fireworks

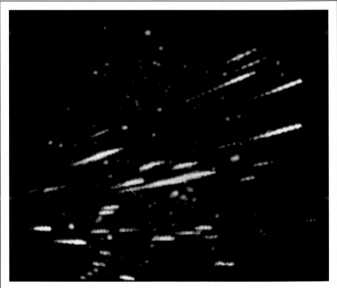

- The Perseid meteor showers resemble a fireworks display in the night sky.

- Comet dust particles ignite the Perseids, as well as most of the meteor showers observed throughout the year.

- Ancient Chinese records,

dated A.D. 36, noted the Perseids and their prolific nature with these words, "More than 100 meteors flew thither in the morning."

- This accounting was the first reportage of what has become the most renowned of all summer meteor showers.

locations and are swift movers. So, it's a good idea to keep your eyes peeled to the sky when observing these summertime meteor showers. The Perseids appear to radiate from the constellation Perseus the Hero. Hence, their name: Perseids.

Often overlooked is the source of these annual meteor showers: the unassuming Comet Swift-Tuttle. This pedestrian comet's dust-laden trail of debris from orbiting the sun cannot help but intersect with our planet's orbit every summer. This "same time next year" occasion spawns a light show that untold people the world over look forward to every July and August.

Summer Meteor Showers

- The Perseids add a spectacular resonance to summertime's warm and reassuring night skies.

- Comet Swift-Tuttle is the celestial body responsible for this annual summer event.

- The Perseid meteor show-

ers are sometimes called the "Tears of Saint Lawrence." Their approximate peak time coincides with the date of Saint Lawrence's death.

- The rate of visible Perseids increases during the predawn hours. This is the case with all meteor showers.

Dark Skies and Meteor Showers

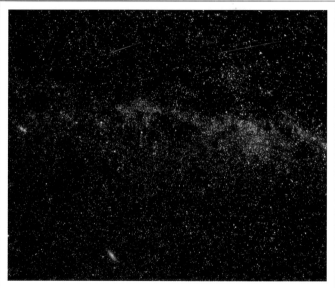

- The best place to observe the Perseids—and all other meteor showers, for that matter—is in a clear dark sky as far away from city lighting as physically possible.

- After midnight and closer to the dawn hours are the best times to view meteor showers.

- As far as stargazing events go, observing meteor showers with the naked eye usually suffices.

- However, weaker showers can often be captured with binoculars and a telescope.

FALL METEOR SHOWERS

Autumn is replete with annual meteor showers: The Orionids, Taurids, Leonids, and Geminids

Fall is a colorful time of year. Beyond the multihued foliage, this season has got meteor showers galore to tout. Beginning in late October and lasting for about a week, the Orionids are active. These meteor showers are something of an encore performance. That is, the second act of Earth's orbit interacting with the rocky residue of Halley's Comet. In the springtime, similar celestial circumstances inspire the Eta Aquarids.

The Taurids are another late October meteor shower. They are active from the last weeks of October into early November. Their radiance is evident in the constellation Taurus the Bull. And because they occur at the end of October, the Taurids are sometimes referred to as "Halloween fireballs." This

Orionids in October

- The Orionids occur in mid- to late October and can be spotted in the constellation Orion.

- The Orionid meteor showers are the second annual event inspired by Halley's Comet. The Eta Aquarids are also Halley's doing and transpire in springtime.

- When at their pinnacles, the Orionids have been known to generate sixty meteors per hour.

- These displays of celestial fireworks can be appreciated with the unaided eye.

Halloween Fireballs

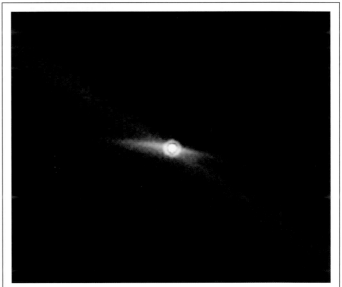

- The Taurid meteor showers annually occur in late October through early November.

- The Taurids appear to radiate from the constellation Taurus the Bull, which is where their name originates.

- The Taurids are not regarded as one of the year's leading meteor showers. Nonetheless, they can be impressive and engender fireballs and bolides (the brightest of fireballs).

- These meteor showers are considered easier to track because they shoot across the sky slower than their peers.

annual meteor shower experience owes Comet Encke a debt of gratitude because this space object's debris footprint makes it all happen.

Mid-November ushers in perhaps the most prolific of all meteor showers: the Leonids. The region to focus on during this time is the constellation Leo, where the meteor showers appear to be streaking through the sky in great numbers. In the past, the Leonids have generated what are called "meteor storms," multiple meteors whizzing across from the sky at the same time.

Last but not least, the Geminids, showering down across the borders of the constellation Gemini, pop onto the scene in mid-December, with December 12 through 14 a typical peak time. The Geminids are at once reliable and prolific meteor showers. Some recent Geminids have spawned 120 to 160 meteors per hour. The Geminids are known to appear in multiple regions of the night sky. First observed a mere 150 years ago, the Geminids owe their existence to an obscure asteroid named "3200 Phaethon." That's kind of how things work in the cosmos. Little things sometimes mean a lot.

Meteor Stormy Night

- Seen in mid-November, the Leonid meteor showers are fast movers.

- Although typically averaging fifteen or so meteors an hour, the Leonids have been extraordinarily prolific at times, generating five hundred and more meteors per hour.

- The Leonids frequently produce fireballs.

- The Leonid meteor showers have generated what are known as "meteor storms," including an 1833 event with an estimated 100,000 meteors or more per hour startling observers.

3200 Phaethon and the Geminids

- The Geminids get their mojo from a celestial body known as "3200 Phaethon." This mysterious object had long been viewed as an asteroid, but now scientists think it might well be the rocky carcass of a former comet.

- The Geminid meteor showers can be spotted in the environs of the constellation Gemini the Twins from early to mid-December.

- The Geminids are among the most reliable meteor showers of the year.

THE ECLIPSE MYSTIQUE
Since recorded time, solar and lunar eclipses have awed human societies

The ancient Greeks applied the word *ekleipsis*, which translates in English as "cease to exist," to a celestial phenomenon that baffled and even frightened them. Indeed, solar and lunar eclipses have long fascinated human cultures. When one outer-space object blocks another one from the view of a third party, we've got an eclipse on our hands.

From our perspective, eclipses involve the sun and the moon. But they are hardly the unique province of Earth, moon, and sun. In fact, they occur throughout the solar system and beyond with countless other celestial objects as the featured players.

As witnessed from Earth's box seat, solar eclipses are

Types of Solar Eclipses

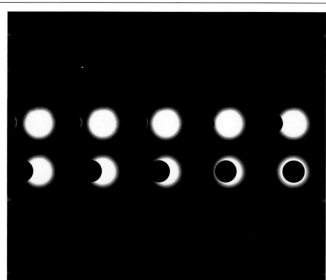

- A solar eclipse sometimes occurs when the moon is closely aligned between our planet and the sun.

- Solar eclipses can take place only during new moon phases, and they are visible only on narrow geographical snippets of Earth.

- During solar eclipses, the moon's shadows—known as the "penumbra" and "umbra"—block sunlight from reaching selected locations.

- Solar eclipses can be total, annular, or partial.

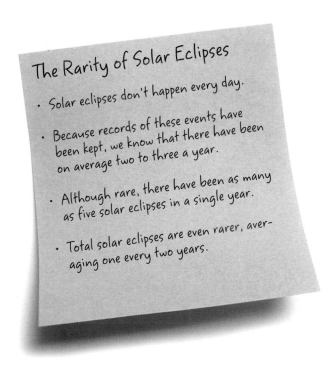

The Rarity of Solar Eclipses

- Solar eclipses don't happen every day.

- Because records of these events have been kept, we know that there have been on average two to three a year.

- Although rare, there have been as many as five solar eclipses in a single year.

- Total solar eclipses are even rarer, averaging one every two years.

certainly more exhilarating than lunar eclipses. When they happen, solar eclipses draw spectators from all parts of the world to witness them. These moments of occultation, as they are called, involve the sun, moon, and Earth in a more or less straight alignment, with the moon blocking some or all of the sun and, of course, the commensurate sunlight. When this scenario plays out in the middle of a sunny day, it can be quite an experience to behold.

As you can imagine, without the knowledge of science that we have today, ancient cultures were understandably alarmed by these turns of events: Earth suddenly and without warning going dark, only to return to light several minutes later. Ancient people reasoned that the gods were at work and relaying some sort of bad tidings.

No ominous message here. Just appreciate when the moon's seeming diameter appears to dwarf the sun and inspires a solar eclipse. Appreciate, too, when the moon sidles into the Earth's shadow and commences a lunar eclipse. These are celestial events to mark on your calendars.

Types of Lunar Eclipses

- Like solar eclipses, lunar eclipses occur when the moon is aligned with our planet and the sun. However, at this time, Earth lies between the moon and the sun.

- Earth's shadows block sunlight from striking the moon during lunar eclipses.

- Whereas solar eclipses occur only during new moon phases, lunar eclipses occur only during full moon phases.

- Lunar eclipses can be either variations of partial or total.

Diamond Ring Effect

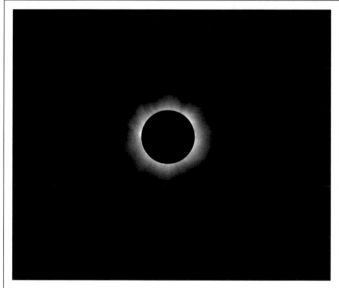

- The diamond ring effect is associated with total solar eclipses.

- Just prior to, and immediately after, the moon's full blanketing of the sun's disk, beads of sunlight are visible around the moon's edges. This is sunlight passing through its jagged and mountainous terrain.

- These beads of light are known as "Baily's Beads" after the man who discovered them: British astronomer Francis Baily.

- When only one remaining bead is visible, it often glows like a diamond ring.

ANNULAR SOLAR ECLIPSE

These eclipses reveal themselves with a gleaming ring around the darkened sun's orb

Solar eclipses are among the most splendid celestial happenings to observe in real time. But like all such recurring events, they are not created equal. There are three types of solar eclipses: total, partial, and annular.

Annular eclipses could best be described as a kind of "close but no cigar" total eclipse. That is, the moon is between Earth and the sun, but its apparent size vis-à-vis the sun's disk is just too small to fully conceal it. What we see during an annular eclipse's peak moment is the darkened orb of the sun surrounded by an extremely radiant ring, a "ring of fire," as it is often called. This makes for quite an impressive visual.

Exactly why does the moon sometimes sustain a total solar

Annular Solar Eclipse

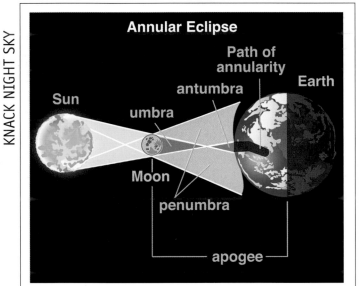

- An annular solar eclipse occurs when the moon is on the far side of its orbit.

- When the sun's outer edges are visible at the eclipse's peak, creating a ring effect, you must be within the moon's antumbra, which extends away from its umbra shadow, to witness it.

- The umbra shadow doesn't reach Earth's surface, but inside what is known as the Path of Annularity—the track of the antumbra— annular solar eclipses occur.

Ring of Fire

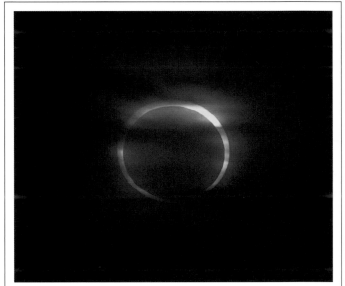

- When the moon cannot completely shroud the sun during an eclipse moment because of distance, the ring of fire effect is visible.

- The ring of fire occurs when the moon is directly in front of the sun.

- From Earth's perspective, a bright ring is discernible around the darkened moon.

- If the moon is on the near side of its orbit during these moments, and closer to the sun, total solar eclipses occur.

eclipse and other times permit a ring of fire to outline an otherwise darkened sun? It's because Earth's only natural satellite orbits its parent planet in elliptical sweeps and not perfect circles. This puts the moon at slightly varying distances from Earth. The differential causes the moon, depending on its exact position, to appear larger or smaller to us. Annular solar eclipses occur when the moon is on the far side of its orbiting trajectory—that is, appearing smaller than the sun at that distance. While on the near side of its orbit, the moon appears larger than the sun and supports a total solar eclipse.

And, yes, although the sun is actually four hundred times larger than the moon, it is also four hundred times farther away from Earth. This accounts for why the moon can sometimes appear bigger than the sun in the sky.

The most spectacular annular solar eclipse of the entire millennium occurred on January 15, 2010. Another solar eclipse combining this type and duration will not come to pass until December 23, 3043.

Annulus

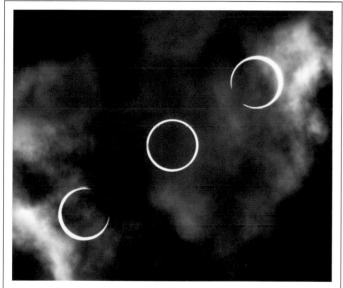

- As rare as total solar eclipses, annular solar eclipses are spectacular to witness.

- During annular eclipses the solar corona is not seen because of the brilliance of what is called the "annulus," the ring around the moon.

- Annular solar eclipses are considered "almost total" solar eclipses.

- Like total solar eclipses, annular solar eclipses have a brief duration: Less than ten minutes is the norm.

Protective Eye Gear

- Protective eye gear is an absolute must to view a solar eclipse.

- There are solar filters, which can be purchased to observe solar eclipses with sunglasses and telescopes.

- It is vital that you know for certain whether these solar filters are the genuine articles.

- Of course, solar eclipses are dangerous for naked-eye viewing. But cameras, binoculars, and telescopes offer no protection. In fact, these magnifying tools of the trade can be even more hazardous to your health.

PARTIAL SOLAR ECLIPSE

These eclipses occur in the moon's outer and faint shadow known as the "penumbra"

Each variation of a solar eclipse occurs when the moon is in its new moon phase. And because the lunar cycle is a shadow shorter than a month's time, it would stand to reason that solar eclipses happen with every new moon. The moon, however, does not orbit Earth in neat circles. Its plane is tilted with respect to our planet's orbit of the sun. This differential is significant enough to cause the moon's shadow to miss our planet most of the time, even during its new moon phases.

The moon's considerable shadow is actually divided into two parts: the penumbra and umbra. The penumbra is its indirect outer shadow that juts off to either side of its umbra. The umbra is the moon's direct inner shadow.

Moon Getting in the Way

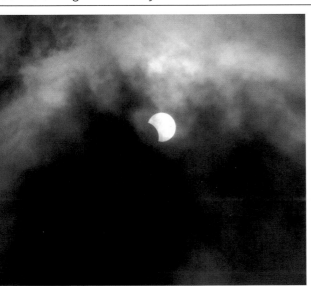

- Partial solar eclipses are the most common.

- The sun appears only partially darkened by the moon's orb.

- They occur when the moon is partially, but not quite fully aligned, between Earth and the sun during its orbit.

- When in this position, the dark side of the moon partly obstructs the sun. This creates a compelling vista for those of us on Earth fortunate enough to be in the moon's shadows.

Partial Solar Eclipse Close-Up

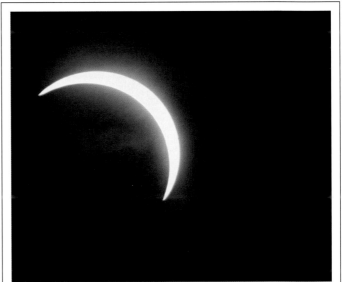

- Partial solar eclipses can occur when the moon's dark and narrow shadow—its umbra—never reaches Earth's surface.

- They also occur when the umbra reaches only a narrow band of territory and the penumbra cuts a significantly wider swath.

- Those viewers located within the umbra enjoy a total or annular solar eclipse, whereas those outside of it—in its more expansive, but lesser shadow—experience a partial solar eclipse.

Partial solar eclipses are seen in the penumbra, in the moon's decidedly fainter and broader shadow. Only portions of the sun are hidden from view in partial solar eclipses. These kinds of solar eclipses are more common and more visible to more people. The penumbra casts a much wider shadow than the umbra, which is where total solar eclipses occur. Keep in mind, too, that even during rare total solar eclipses, most witnesses are seeing only a partial. Solar eclipses are not seen throughout the entire world but rather only in a small radius.

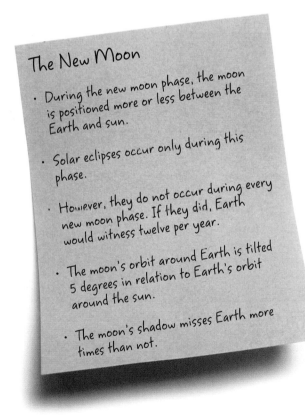

The New Moon

- During the new moon phase, the moon is positioned more or less between the Earth and sun.

- Solar eclipses occur only during this phase.

- However, they do not occur during every new moon phase. If they did, Earth would witness twelve per year.

- The moon's orbit around Earth is tilted 5 degrees in relation to Earth's orbit around the sun.

- The moon's shadow misses Earth more times than not.

····················· RED ● LIGHT ··············

Except during the peak moment of a total solar eclipse, when the sun is completely hidden from view, solar eclipses are dangerous to look at without appropriate solar filters. Sunglasses do not ensure safe viewing of an eclipse. Looking through binoculars, telescopes, and cameras without proper solar filters is equally dangerous. Permanent retinal damage can occur in a flash.

Penumbral Shadow

- Those viewers on Earth experiencing this magnificent partial solar eclipse are within the moon's penumbral shadow.

- This is the moon's more sprawling and dimmer shadow.

- All types of solar eclipses span limited areas. That is, solar eclipses are not worldwide events, or even hemispheric events, but rather confined to regions where the shadows are cast.

- Lunar eclipses, on the other hand, are observable to one and all with a visible moon in the sky.

ECLIPSES

TOTAL SOLAR ECLIPSE

Total solar eclipses occur in the moon's inner and dark shadow known as the "umbra"

Total solar eclipses are the ultimate eclipse events. They don't occur as frequently as other eclipses, and their "period of totality" widely varies. That is, the length of time the sun remains wholly concealed behind the moon's more predominant disk. This is, of course, from the perspective of Earth, where closer objects, like the moon, can obscure much farther and significantly larger objects like the sun.

All solar eclipses are limited to a small segment of Earth caught in the moon's shadow. In order to witness a total solar eclipse, it's imperative that you be within the portion of the moon's inner and dark shadow known as the "umbra."

The moon's umbra can span 10,000 miles (16,000 kilometers)

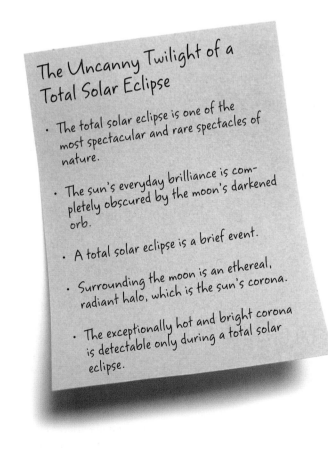

The Uncanny Twilight of a Total Solar Eclipse

- The total solar eclipse is one of the most spectacular and rare spectacles of nature.

- The sun's everyday brilliance is completely obscured by the moon's darkened orb.

- A total solar eclipse is a brief event.

- Surrounding the moon is an ethereal, radiant halo, which is the sun's corona.

- The exceptionally hot and bright corona is detectable only during a total solar eclipse.

Corona on Display

- The total solar eclipse exposes the sun's corona for all to see.

- The solar corona—the sun's outer atmosphere—is incredibly and mystifyingly hot. And it is in evidence only during the few minutes of total solar eclipses.

- Scientists are uncertain why the solar corona, when it is farther away for the sun's energy source, registers higher temperatures than the chromosphere below it.

- The solar corona is extremely dangerous to look at without appropriate eye protection.

but is rarely longer than 100 miles (160 kilometers). This means that when its dark shadow reaches Earth on eclipse occasions—with the moon is between the sun and Earth—there is a relatively small area of the planet's surface bearing witness to the eclipse. On July 22, 2009, a total solar eclipse was visible over much of the continent of Asia. It established a record for the fledgling twenty-first century with a period of totality, in certain locations, of 6 minutes and 39 seconds.

During a total solar eclipse, the sun's chromosphere briefly reveals itself as a red glow—a thin ring—encircling the obscured photosphere. The sun's photosphere is dangerous to look at, but the chromosphere is doubly so because it is actually hotter and brighter.

Although highly anticipated, total solar eclipses come and go in a matter of minutes. With Earth, moon, and the sun in perpetual motion, stasis is never possible.

Path of Totality

- The Hinode satellite captures a captivating image of a total solar eclipse.

- Only when the moon's umbral shadow—its inner and darker shadow—brushes Earth can we experience a total solar eclipse.

- The umbral shadow's trail across the surface of Earth is called the "path of totality."

- Typically, when this dark shadow touches Earth's surface, only 1 percent or less of the landmass is impacted. You would have to be within that slight percentile to witness a total solar eclipse.

Baily's Beads

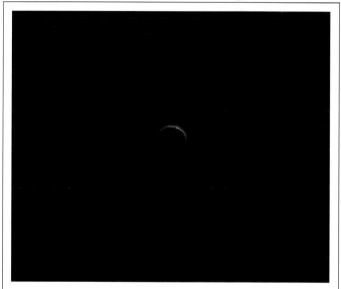

- Baily's Beads are on full display during the moments before a total solar eclipse.

- The craggy lunar topography adds special effects to a spectacular and rare total solar eclipse.

- Baily's Beads reveal themselves as sunlight finds its way through the moon's uneven terrain just before—and then again just after—the total eclipse of the sun.

- These beads of light supply icing on the cake of a celestial event of epic proportions involving Earth, the moon, and sun.

PARTIAL LUNAR ECLIPSE

Earth's shadow blocks some sunlight from reaching the moon during a partial lunar eclipse

Like solar eclipses, lunar eclipses involve the Earth, moon, and sun. These celestial bodies are again neatly aligned in their orbits. But this time, the moon isn't the centerpiece between Earth and the sun. Earth has assumed that position, with the moon directly behind it. In this scenario, the sunlight blocker is Earth, and the moon is sometimes cast into its shadows.

A partial lunar eclipse occurs when Earth's shadow blocks some, but not all, sunlight from reaching the moon's surface. Quite unlike solar eclipses, lunar eclipses can last for more than three hours. Solar eclipses are over and done with in minutes. And lunar eclipses are observable to everybody and anybody enjoying nighttime. In other words, half the planet

Sun Blocker

- This partial lunar eclipse is the consequence of Earth blocking the sun's rays from bathing the moon.

- The moon is situated behind the Earth in its orbit. Earth resides in the middle between the sun and the moon.

- Lunar eclipses occur only during full moon phases. This is the opposite of solar eclipses, which occur only during new moon phases.

- Whereas a solar eclipse lasts only several minutes, a lunar eclipse can last several hours.

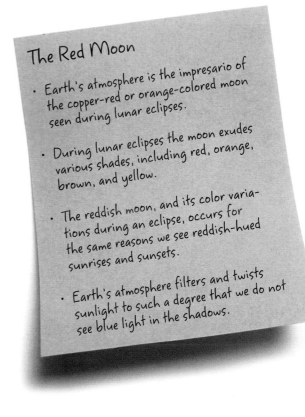

The Red Moon

- Earth's atmosphere is the impresario of the copper-red or orange-colored moon seen during lunar eclipses.

- During lunar eclipses the moon exudes various shades, including red, orange, brown, and yellow.

- The reddish moon, and its color variations during an eclipse, occurs for the same reasons we see reddish-hued sunrises and sunsets.

- Earth's atmosphere filters and twists sunlight to such a degree that we do not see blue light in the shadows.

can check out a lunar eclipse. Solar eclipses are available to only select parts of the planet.

Two or more partial lunar eclipses occur each year. Total lunar eclipses are harder to come by. Lunar eclipses occur only during full moons. When the moon is positioned behind Earth, it is in this phase. So, as you can imagine, a full moon getting eclipsed by Earth's shadow is a phenomenon worth witnessing.

In the Celestial Shadows

- This partial lunar eclipse reveals a portion of the moon passing through Earth's umbral shadow.

- The umbral shadow is the dark inner section of our planet's shadow.

- The distance between the Earth and moon at the time of the eclipse affects its duration.

- The moon's elliptical orbit ensures that the distance between it and Earth varies. Its average distance away is 238,000 miles (383,000 kilometers).

Unfolding Eclipse

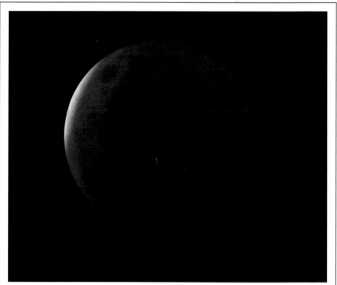

- A partial lunar eclipse adds a special touch to the night sky.

- A partial lunar eclipse can last two to three hours or more, supplying observers with ample time to witness a remarkably rare celestial event.

- During its orbit, the moon's motion is slowest when near its apogee vis-à-vis Earth. This celestial languor adds to the duration of lunar eclipses.

- Unlike solar eclipses, lunar eclipses can be observed with the naked eye without worry.

TOTAL LUNAR ECLIPSE

A total lunar eclipse, sometimes called the "red moon," is a celestial sight to behold

Although partial lunar eclipses are interesting enough, they supply different visuals from total lunar eclipses, which are decidedly rarer. A total lunar eclipse occurs when the moon completely ventures into Earth's umbra. This is Earth's inner and dark shadow, which blankets the entire moon at this time. But fear not, this eclipse image isn't utter darkness.

After all, it's nighttime. What kind of sighting would that be? Rather, the moon appears darkened—yes—but visible in its entirety. Indirect sunlight finds its way through the omnipresent shadow. In fact, the moon appears almost reddish in color. This totality is often called a "blood moon" because of the moon's distinctive shade during a total lunar eclipse.

In Earth's Umbral Shadow

- The complete moon is passing through Earth's umbral shadow, producing a total lunar eclipse. The moon's copper-red color is impressive.

- Approximately one-third of all lunar eclipses are of the penumbral variety, which cannot be detected with the naked eye.

- Earth's penumbral shadow is a region where some, but not all, of the sun's penetrating rays and lighting are blocked from reaching the moon.

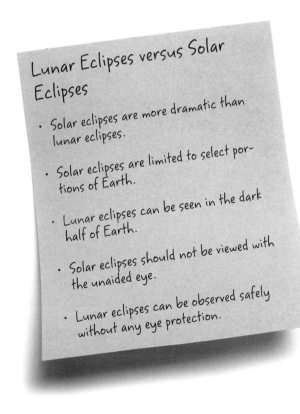

Lunar Eclipses versus Solar Eclipses

- Solar eclipses are more dramatic than lunar eclipses.

- Solar eclipses are limited to select portions of Earth.

- Lunar eclipses can be seen in the dark half of Earth.

- Solar eclipses should not be viewed with the unaided eye.

- Lunar eclipses can be observed safely without any eye protection.

The reddish hue is courtesy of Earth's complex layers of atmosphere. They filter out the sunlight's blue shades and leave behind only a reddish-orange amalgam. The same filtering effect occurs during sunrises and sunsets, when the sun appears red and orange in color. Its literal color scheme doesn't change throughout the day, but when the sun's closer to the horizon it passes through more atmosphere. And just like in a total lunar eclipse, it sifts out particular colors and leaves others behind. Indeed, without Earth's atmosphere on the job, a total lunar eclipse would be a total blackout.

Unfurling Total Lunar Eclipse

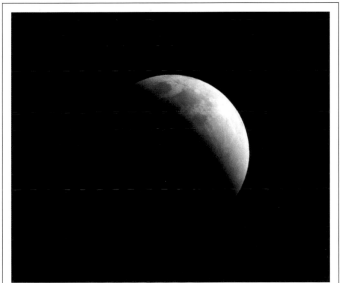

- A partial phase of a total lunar eclipse is an intriguing spectacle.

- Early twentieth-century astronomer André Louis Danjon fashioned what is now known as the "Danjon scale," which scores the brightness of lunar eclipses.

- A score of zero on his scale means that there is a dark umbra and that you barely, if at all, see the moon.

- The highest score of 4 on his scale reveals a bright reddish or orange umbra with a blue-tinted rim.

Almost There

- The most mesmerizing aspect of a total lunar eclipse is the moon's color scheme as seen from Earth.

- Earth's atmospheric layers paint a portrait of the moon with multiple hues plucked off the celestial palette.

- Without Earth's atmosphere between us and our only natural satellite, the moon would be completely black during a lunar eclipse.

- Instead we see colors ranging from dark brown to yellow to orange to bright red.

SPACE WEATHER

Our weather occurs in the troposphere; multicolored space weather occurs in the thermosphere

Earth's immediate and densest atmospheric layer—the troposphere—extends from the planet's surface almost 10 miles up. It is the region where the perpetually changing events that we call "weather" play out.

At the uppermost end of Earth's atmosphere, known as the "thermosphere," which shares space with the highly charged ionosphere, another brand of ever-changing weather called "space weather" occurs. In sharp contrast to what we experience here at the surface, this weather phenomenon arises in a layer of our atmosphere that is very thin. Solar radiation—that is, solar wind—stirs up energy particles here in this near-space region of Earth's atmosphere.

Solar Flare in Ultraviolet Light

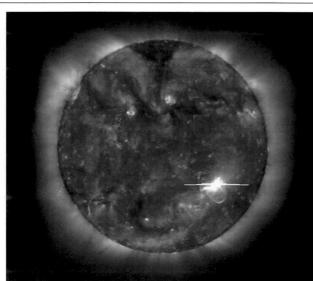

- Using false color imagery, NASA's Solar and Heliospheric Observatory (SOHO) permits us a rare glimpse at a potent solar flare erupting in the sun's atmosphere.

- The layers of churning gases that make up the sun occasionally explode, sending solar flares into outer space.

- These events occur when mounting magnetic energy in the solar atmosphere is unleashed.

- Solar flares snake through the corridors of interplanetary space.

Earth's Magnetosphere

- Earth sports the strongest magnetosphere of all the terrestrial planets.

- The magnetic field contains north and south poles.

- Earth's magnetic field thwarts the most deleterious impact of solar wind.

- Energized particles from the solar wind that infiltrate the magnetosphere create aurorae.

Essentially, atoms and molecules respond to an endless barrage of particles entering the atmosphere. That is, they become "excited." And excited atoms and molecules can calm themselves down only by returning to nonexcited states—their ground states. They accomplish this by emitting photons, which produce light. The careening and multihued light curtains sometimes visible in the upper atmosphere are known as "aurorae" (singular: aurora).

YELLOW LIGHT

If you plan on making a trip to more hospitable locales to view the Northern Lights or Southern Lights, check out what the sun's been up to. The sun regularly emits solar flares caused by explosions within its atmosphere. These flares contribute to the intensity of aurorae. The more solar flare activity, the more dynamic the space weather.

Solar Flares Do Their Thing

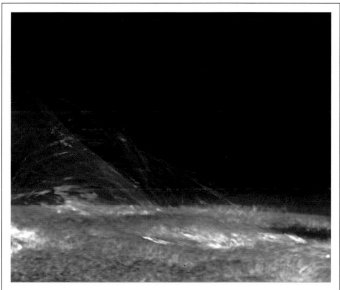

- Solar flares in the regions of the sun's atmosphere are observed when unexpected and perceptible upticks in brightness occur.

- Solar flares discharge tremendous amounts of energy into the nooks and crannies of our solar system.

- The solar wind carries the flare's highly energized particles throughout the celestial ether.

- Upon entering Earth's magnetosphere, the solar flares influence what is known as space weather.

Looking Down on Aurora Australis

- As seen from outer space, the Aurora Australis is visible over Antarctica.

- Courtesy of energized particles deposited by the solar wind, colorful aurorae are generated in Earth's upper atmosphere.

- Aurorae have been likened to neon lights. That is, the colors emitted are the consequence of different gases getting overly excited.

- Excited atoms and molecules store energy. They return to their nonexcited ground states by emitting photons—that is, light.

AURORA COLORS

The aurora colors are different gases at different altitudes interacting with energy particles

The multiple colors associated with aurorae—red, green, yellow, blue, and purple—are what make these atmospheric light shows worth the price of admission. They are never the same in intensity or color schemes. No two aurorae are ever alike. Some appear faint and short-lived, whereas others are bright and blanket the night sky. What exactly is going on with these color variations? What is causing these pulsating rainbows in the night skies? It again all redounds to particles—electrons and protons—from outer space entering our upper atmosphere and interacting with, and exciting, atoms and molecules. That is, solar radiation meeting and greeting the variety of gases contained in our atmospheric bubble. And

KNACK NIGHT SKY

Green Aurora

- An aurora unfurls an eye-catching green curtain across the night sky.

- Earth's space weather is active as high-energy electrons from the planet's magnetosphere interact with atoms and molecules.

- The green color indicates oxygen atoms, at approximate atmospheric altitudes of 60 miles (100 kilometers) to 150 miles (250 kilometers), as the responsible party.

- Green light is released as the oxygen atoms return to their unexcited, ground states.

Red Aurora

- A spectacular red aurora cascades in the night sky, supplying fortunate observers with a light show to remember.

- Red-colored aurorae are the most atypical of these light shows.

- Similar to green aurorae, the red color is indicative of oxygen atoms at work. They are at once getting excited and returning to their unexcited states but at even higher levels in the atmosphere.

- Aurorae colors are the individual signatures of atoms and molecules.

when in these highly excited states, some species of atoms and molecules emit one color to stabilize, whereas others release another color to calm down.

Oxygen atoms, for instance, are wont to flaunt reds at the highest altitudes. That is, at levels beyond 180 miles (300 kilometers) into the atmosphere. From our vantage point, the red often appears as a more brownish-red. Red aurorae are the rarest. At lower altitudes, the same oxygen gas lets loose greens, yellows, and oranges. The presence of nitrogen frequently delivers colors like blue and purple.

Greenish-Blue Aurora

- A bluish aurora light show indicates the presence of ionic nitrogen, which releases a warm-colored illumination into the night sky.

- Whereas oxygen atoms ordinarily release photons of green and red, nitrogen exudes the color blue.

- Photographs of aurorae are often more vividly colored than observations with the naked eye. So, don't be surprised when you develop your film to see an even more stunning aurora.

- Quality film is frequently more sensitive to colors.

Purple-Violet Aurora

- The color purple swathes the night sky as an added bonus for aurorae aficionados.

- Nitrogen atoms are again the foundation of this particular light show.

- Nitrogen gases can emit purple, blue, and violet hues.

- Nitrogen molecules are also responsible for reddish-purple aurorae colors sometimes seen on their rippled edges. This color combination is relatively rare.

AURORAS

219

AURORA BOREALIS

The Northern Lights fill the night skies with brilliant curtains of light

The Aurora Borealis, also known as the "Northern Lights" or "Northern Polar Lights," are extraordinary pulsating light shows in the upper regions of Earth's atmosphere. They are visible to us on the surface but mostly in the far northern latitudes. When intense and prolific, Aurora Borealis take their light show to more places. They have actually been spotted as far south as Florida.

Aurora Borealis frequently cascade across the sky, twisting right and then left while generating vivid curtains of light. The colors are sometimes uniform red, green, yellow, blue, or purple but at other times an intoxicating blend of the color spectrum that will often last ten and fifteen minutes before vanishing altogether.

The Aurora Borealis received their name from the Roman goddess of dawn, Aurora, and the Greek word for "north wind": *boreas*. Typically these light shows are most active

Aurora Borealis over Alaska

- From the catbird's seat in frigid Alaska, the Aurora Borealis (the Northern Lights) do what they do best.

- The Aurora Borealis are most often seen in close proximity to the North Pole.

- This is due to the machina-tions of Earth's magnetic field and, of course, the longer nights.

- To increase your chances of witnessing this space weather, the night sky should be both clear and dark. Many aurorae are too faint to see with excessive cloud cover or light.

Atmosphere Hijinks

- The Earth's magnetosphere and solar wind join forces and spawn a greenish Aurora Borealis.

- The Aurora Borealis appear to observers as vibrant and vibrating curtains draped across the night sky.

- This visual is the conse-quence of electrons twist-ing and turning along our planet's magnetosphere, colliding with gases in the atmosphere, and remaining on or near the field lines.

- Greater amounts of energy in the electrons engender more prolific Aurora Borea-lis shows.

during maximum stretches of the sun's solar cycle. During the maximum cycle, sunspots on the sun's surface trigger explosions and superhot solar flares, which ride the solar wind through space, ultimately barreling into the Earth's upper atmosphere and its magnetic field known as the "magnetosphere." The result of this solar radiation encountering the myriad gases of our atmosphere is a picture show that must be seen to be believed.

ZOOM

Although aurorae displays are quite magnificent to witness, they sometimes cause a few problems on the surface. Electric power lines have been negatively affected by aurorae with resultant blackouts. Short-wave communications have been disrupted. And meteorological satellites measuring our weather have been completely disabled by this space weather.

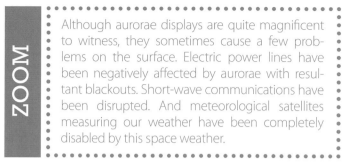

Aurora Borealis and Wintertime

- Dazzling Aurora Borealis paint a portrait in the sky worthy of a frame in a museum.

- Although total darkness is considered premium aurora-viewing time, twilight aurorae are often quite compelling. Bright aurorae can even coexist with the moon.

- In fact, for photographing the Aurora Borealis, a moon in the night sky can be beneficial. You don't need exposures nearly as long with the moon's generous light source helping out.

Aurora Borealis in Twilight

- The Aurora Borealis in the twilight welcome in the evening with celestial eye candy.

- Generally speaking, the prime time for viewing the Northern Lights is around midnight.

- However, these impressive light shows can occur at all hours. They become visible when the sky is dark or transitioning into darkness.

- The Aurora Borealis occur just about all the time but increase in frequency and brilliance during periods of greater solar activity.

AURORAS

AURORA AUSTRALIS

The Southern Lights distribute sheets of vibrating colors into the night skies

At both our planet's poles, the upper atmosphere is ground zero for the mesmerizing spectacle known as "aurorae." In the Northern Hemisphere, these undulating displays of multicolored light in the night sky are called "Aurora Borealis." In the Southern Hemisphere, they have been christened "Aurora Australis." *Australis* is a Latin term for "of the south."

Although Aurora Borealis (the Northern Lights) are much better known than Aurora Australis (the Southern Lights), it is not because their respective light shows are noticeably different. Aurora Australis are equally impressive. It's just that fewer people experience them. It is a matter of geography.

The Aurora Australis originate at the South Pole, where

Southern Hemisphere's Aurora

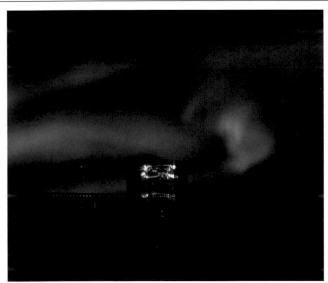

- The Aurora Australis are the Southern Hemisphere's equally impressive space weather.

- The Southern Lights dance across the night sky in latitudes near the South Pole. They appear in various colors and guises.

- These light shows are akin to the Aurora Borealis, only visible on the other side of the planet.

- Appearing as twisting curtains of light, semicircles, and streamers are par for the aurorae course.

Aurora Australis at the South Pole

- The South Pole is ground zero for the Aurora Australis.

- Electrons mosey along Earth's magnetic field lines, which inspire the Aurora Australis.

- At the two poles, Earth's magnetic field lines are

jam-packed, which is why aurorae are prominent in the neighborhoods of both the North Pole and the South Pole.

- The Aurora Borealis are more often observed than the Aurora Australis because of the respective stages they play on.

Earth's magnetic field is—along with the North Pole—at its most persuasive. So, this spectacular demonstration of space weather's epicenter is located over the continent of Antarctica. From October to through February—spring and summer months in the Southern Hemisphere—this sprawling region of cold and ice is bathed in almost perpetual sunlight. Naturally, twenty-four hours of sunshine make seeing the Aurora Australis well nigh impossible.

Curtains of Light

- The Aurora Australis's flowing lights undulate on the grandest stage of all.

- It has been analogized that our viewing aurorae from Earth's perch is akin to a colony of ants staring up at colorful kitchen window curtains.

- Whether or not space weather comes with audio is still the subject of scientific debate.

- If you believe you heard the Aurora Australis or Aurora Borealis emit sounds, report it.

Outer-Space Perspective

- NASA's *Spacelab 3* is treated to the Aurora Australis from an outer-space perspective.

- Although aurorae are sights for sore eyes, they are nonetheless caused by highly energized particles blowing in the solar wind.

- Altered behavior in air and marine navigational tools have been blamed on the progenitors of colorful aurorae.

- When observing the Northern or Southern Lights, consider too what's transpiring in the atmosphere.

AURORAS

NORTHERN LIGHTS VIEWING

The closer to the North Pole, the better are the odds of witnessing the Aurora Borealis

Experiencing the Aurora Borealis is not as straightforward as checking out a lunar eclipse or a comet's expected appearance in the night skies. No, for these recurring phenomena, you optimally have to be within a prescribed oval band: one with its epicenter at the North Magnetic Pole and with a width that ranges from 6 to 600 miles (10 to 1,000 kilometers). Near this

band, when aurorae are active, you are on what is considered prime terra firma to witness nature's greatest light show. This oval, however, does not conform to uniform latitudes. That is, East Coast denizens of the United States are more apt to witness the Aurora Borealis than are people on the West Coast.

Some of the ideal spots for viewing the Aurora Borealis

Aurora Borealis in Cold Climes

- The Aurora Borealis loom large over frigid of Alaska.

- Fairbanks, the capital of the largest U.S. state, is a prime location to witness the Aurora Borealis.

- If you want to increase your chances of seeing the Northern Lights, it is imperative that you venture to the highest northern latitudes.

- During maximum auroral events, which are the exception to the rule, the Northern Lights can be seen in much lower altitudes than the norm.

Northern Lights 101

- The Northern Lights are evident throughout the year.

- Space weather occurs night and day.

- Most aurorae are not visible during the daytime.

- There are active and inactive periods of aurorae that correspond with solar activity.

- Earth weather plays a significant part in viewing, and not viewing, the Aurora Borealis.

include northern Alaska (Fairbanks, for example, the state's capital), northern Greenland, northern sections of Russia, and the Scandinavian countries of Europe. It should be pointed out that each location above the frigid Arctic Circle is bathed in sunlight all day long from April through September—that is, spring and summer in the Northern Hemisphere. And the Northern Lights are almost exclusively night shows. Wintertime would be best for viewing the Aurora Borealis in these places. The liveliest and most dazzling light shows often occur near the midnight hour. Prime aurorae viewing hours are between 11 p.m. and 2 a.m.

•••••••••••••• RED ● LIGHT ••••••••••••••

Planning an Aurora Borealis hunting trip next summer? No matter your location, summertime means longer days and less darkness. Great for enjoying many vacation activities, sure, but not—alas—for seeing aurorae. Springtime, near the equinox, flaunts a proven track record of winning appearances. The nights are still relatively cold—with minimal humidity and pollution—furnishing you with fine views of the night skies.

Aurora Borealis Venture South

- The Aurora Borealis venture down into the night skies of Wisconsin.

- Although aurorae sightings are not regular occurrences this far south, the Northern Lights do on occasion venture into the lower forty-eight states.

- The Aurora Borealis have been spotted in many areas in Canada, as well as states on the Canadian border, including Montana, Minnesota, Michigan, and North Dakota.

- Getting away from the bright lights of big cities is always sound advice.

Aurora Borealis Over Finland

- The skies of Finland are awash in the rolling green of the Aurora Borealis.

- There are five Scandinavian countries: Denmark, Finland, Iceland, Sweden, and Norway.

- The polar regions of Finland, Iceland, Sweden, and Norway have long track records of recurring and spectacular Aurora Borealis sightings.

- The cold, crisp, and clean air of these countries in high northern latitudes is the perfect recipe for the Aurora Borealis.

AURORAS

SOUTHERN LIGHTS VIEWING

The nearer to the South Pole, the better are the chances of seeing the Aurora Australis

Just as is the case with their polar opposite light shows to the north, the Aurora Borealis, a similar playbook applies to the Aurora Australis. But with the Aurora Australis, the script is turned upside down.

Fundamentally, to view the Aurora Australis, the Southern Lights, you, too, should be within an explicit oval band. This will increase your chances of not only witnessing the Aurora Australis but also of seeing them at their finest. The Southern Lights' band is centered at the South Magnetic Pole. Its width ranges from 6 to 600 miles (10 to 1,000 kilometers). Again, this is the prime terrain for seeing these aurorae in action. Regions of the world known for splendid Aurora Australis sightings include, of

Aurora Australis in the Antarctic

- The Aurora Australis once again thaw out the Antarctic night sky with a colorful display of lights.

- Although these are much less observed than their northern counterparts, they are as impressive as the Aurora Borealis.

- The Antarctic weather is rather severe and not exactly a hotspot for tourists.

- During the months of June, July, August, and September—the Antarctic winter—the Southern Lights perform mainly for the benefit of penguins.

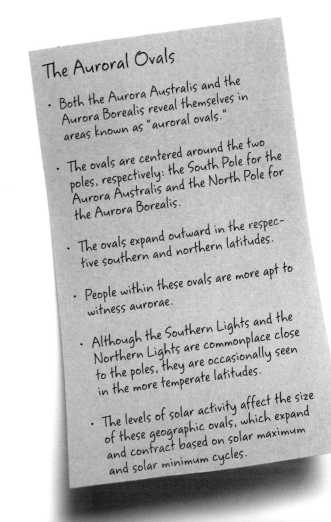

The Auroral Ovals

- Both the Aurora Australis and the Aurora Borealis reveal themselves in areas known as "auroral ovals."

- The ovals are centered around the two poles, respectively: the South Pole for the Aurora Australis and the North Pole for the Aurora Borealis.

- The ovals expand outward in the respective southern and northern latitudes.

- People within these ovals are more apt to witness aurorae.

- Although the Southern Lights and the Northern Lights are commonplace close to the poles, they are occasionally seen in the more temperate latitudes.

- The levels of solar activity affect the size of these geographic ovals, which expand and contract based on solar maximum and solar minimum cycles.

course, Antarctica, ground zero, but also parts of Australia, New Zealand, and southern points in South America.

But, as is always the case with aurorae, there are no guarantees that you will see them—in the geographic oval or outside of it. Countless factors are always at play. Foremost, you need geomagnetic disturbances on your side. That is, hypersolar activity greasing the skids and commingling with our upper atmosphere and magnetic fields. Without an active sun and highly charged solar wind, there are fewer aurorae events to speak of.

Aurora Australis over Tasmanian Skies

- The Aurora Australis over the Tasmanian night skies are not an unusual occurrence.

- The Southern Lights are as prolific as the Northern Lights. The dearth of photos and information on the Aurora Australis, as compared with the Aurora Borealis, is solely based on population and access to viewing.

- The Northern Lights are definitely more reachable than the Southern Lights.

- There is significantly more landmass in northern polar regions than in southern polar regions.

Active Aurora Australis

- The Southern Lights colorfully radiate and ripple over the night skies of Australia.

- Before the Aurora Australis can be seen in locations beyond the vastness of the Antarctic desert, the aurorae must be extremely active.

- The large continent of Antarctica is surrounded by ocean waters and receives the lion's share of Aurora Australis sightings.

- During active periods, the auroral oval expands into places like Tasmania and the southern reaches of New Zealand.

AURORAS

RESOURCES

Astronomy is the one branch of science that puts out a welcome mat for amateurs. That is, no university degree or other specialized training is required to survey the night skies. And there are abundant astronomical resources at your fingertips, too: associations to join, informational Web sites to further your education, and publications to peruse to keep you abreast of breaking cosmic news. In addition, specialty retailers offer the tools of the stargazing trade. From binoculars to telescopes to sky maps, countless astral products are available in the marketplace. In fact, many of the hobby's resources have the amateur astronomer in mind. No matter the level of your stargazing IQ, there is something for everyone—beginners, seasoned veterans, and professional astronomers, too—in the eclectic resources listed here.

Further Education

Astronomer Organizations

American Association of Amateur Astronomers (AAAA)
P.O. Box 7981
Dallas, TX 75209-0981
www.astromax.com

American Meteor Society
www.amsmeteors.org

Astronomical League
www.astroleague.org

Astronomical Society of the Pacific
390 Ashton Ave.
San Francisco, CA 94112
(415) 337-1100
www.astrosociety.org

International Astronomical Union (IAU)
www.iau.org

Celestial Calendars

EarthSky
www.earthsky.org/tonight/home

Faulkes Telescope Project
www.faulkes-telescope.com

Sea and Sky
www.seasky.org

StarDate Online
StarDate
1 University Station A2100
Austin, TX 78712
(512) 471-5285
www.stardate.org

Informational Web Sites

Astronomy Today
www.astronomytoday.com

Catching the Light
www.astropix.com

Space.com
www.space.com

Stargazing for Everyone
www.stargazingforeveryone.com

Universe Today
www.universetoday.com

NASA

NASA's Astronomy Picture of the Day
antwrp.gsfc.nasa.gov/apod
Public Communications Office
NASA Headquarters
Suite 5K39
Washington, DC 20546-0001
(202) 358-0001
www.nasa.gov

Gear

Astrophotography and Solar Filters

Lumicon
750 E. Easy St.
Simi Valley, CA 93065
(800) 420-0255
www.lumicon.com

Thousand Oaks Optical
(928) 692-8903
www.thousandoaksoptical.com

Binoculars and Binocular Chairs

Astrogizmos.com
(540) 898-0636
www.astrogizmos.com

BigBinoculars.com
Oberwerk Corporation
2790-C Indian Ripple Rd.
Beavercreek, OH 45440
(866) 244-2460
www.bigbinoculars.com

BinocularsWorldwide.com
(800) 331-6595
www.binocularsworldwide.com

Sky Maps and Accessories

Skymaps.com
www.skymaps.com

Starry Night Store
89 Hangar Way
Watsonville, CA 95076
(800) 252-5417
www.starrynightstore.com

Space Souvenirs

Kennedy Space Center Gift Shop
(800) 621-9826
www.thespaceshop.com

Universe Collection
(203) 393-3395
www.universecollection.com

Telescopes

OpticsPlanet.com
(800) 504-5897
www.opticsplanet.net

Orion Telescopes & Binoculars
(800) 447-1001
www.telescope.com

Telescopes.com
(800) 303-5873
www.telescopes.com

Publications

Magazines

Amateur Astronomy Magazine
511 Derby Downs
Lebanon, TN 37087
(615) 332-5555
www.amateurastronomy.com

Astronomy
Kalmbach Publishing Co.
21027 Crossroads Circle
P.O. Box 1612
Waukesha, WI 53187-1612
(800) 533-6644
www.astronomy.com

Night Sky Observer
www.nightskyobserver.com

Sky & Telescope
90 Sherman St.
Cambridge, MA 02140
(800) 644-1377
www.skyandtelescope.com

Sky News
(866) SKY-0005
www.skynewsmagazine.com

Books

The 100 Best Astrophotography Targets
Ruben Kier
Springer, 2009

Astronomy: A Self-Teaching Guide
Dinah L. Moche
Wiley, 7th edition, 2009

The Backyard Astronomer's Guide
Terrence Dickinson and Alan Dyer
Firefly Books, 3rd edition, 2008

Nightwatch
Terrence Dickinson
Firefly Books, 4th edition, 2006

Sky & Telescope's Pocket Sky Atlas
Roger W. Sinnot
Sky Publishing, 2006

Star Watch
Philip S. Harrington
Wiley, 2003

Astronomy Aids

Eclipses

Mr.Eclipse.com
www.mreclipse.com

NASA Eclipse Web Site
http://eclipse.gsfc.nasa.gov/eclipse.html

Planetarium Software

RedShift7
www.redshift-live.com

SkyMapPro
www.skymap.com

SkyVoyager
(800) 493-8555
www.carinasoft.com

TheSky6
www.bisque.com

Planets

NASA's Welcome to the Planets
http://pds.jpl.nasa.gov/planets/

Smithsonian National Air and Space Museum: Exploring the Planets
www.nasm.si.edu/etp/

Solar System

NASA's Solar System Simulator
http://space.jpl.nasa.gov/

Views of the Solar System
www.solarviews.com

Interactive Astronomy

Interactive Outer Space

GoogleSky
www.google.com/sky/

Sky-Map.org
www.sky-map.org

Networking and Community

KidsAstronomy.com
www.kidsastronomy.com

NASA's Imagine the Universe!
http://imagine.gsfc.nasa.gov

NASA's Night Sky Network
http://nightsky.jpl.nasa.gov/

Windows to the Universe
www.windows.ucar.edu/

Observatories

Smithsonian Astrophysical Observatory (SAO)
Harvard-Smithsonian Center for Astrophysics
60 Garden St.

Cambridge, MA 02138
(617) 495-7463
www.cfa.harvard.edu/sao/

Solar and Heliospheric Observation (SOHO)
http://sohowww.nascom.nasa.gov/

Space Cams and Multimedia

Kennedy Space Center Video Feeds
http://science.ksc.nasa.gov/shuttle/countdown/video/

NASA Multimedia
www.nasa.gov/multimedia/

Space.com Cams
www.space.com/spacewatch/cams.html

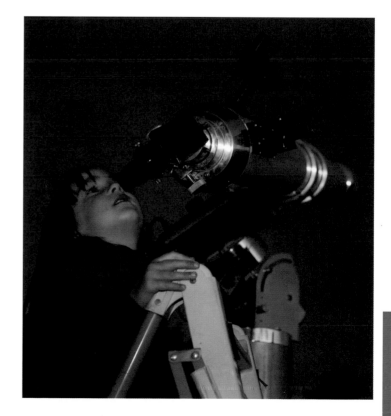

GLOSSARY

Asteroid: A piece of a larger celestial object of uncertain origin—often presumed to be the remains of a planet that formed billions of years ago—that presently orbits the sun or a planet.

Astrophotography: Specialized photography of celestial bodies, including the sun, moon, stars, planets, and deep-sky objects.

Aurora: Colorful bands of light in the night sky inspired by charged particles from outer space colliding with the myriad gases in Earth's upper atmosphere.

Binary Star System: A pair of stars that orbits a shared center of mass and often appears as a single star to observers.

Black Hole: An area in space created by the collapse of a large star where the gravitational forces are so extreme that no matter or energy—not even light—can ever escape it.

Catadioptric Telescope: A compact telescope that combines elements of both refractor and reflector telescopes.

Celestial Sphere: An imaginary depiction of the night sky that enables astronomers to appreciate the role of Earth's rotation of space's myriad objects as well as identify their positions at any given time.

Comet: An irregular space body consisting of a unique blend of frozen gases, dust, and solid materials—volatile and nonvolatile alike—that orbits the sun.

Constellations: Arbitrary demarcations and names applied to patterns of stars and specific regions in the night sky.

Deep-sky Objects: Celestial bodies outside of our solar system including nebulae and star clusters.

Diurnal Motion: The apparent daily movement east to west—caused by Earth's rotation—of most stars on the celestial sphere.

Eclipse: A partial or total concealment of one celestial body by another.

Ecliptic: The sun's apparent movement—in concert with Earth orbit—along the celestial sphere throughout the year.

Galaxy: A distinctive grouping of stars and interstellar material bound together by a common center of gravity.

Gas Giant Planets: Planets in our solar system consisting of most gases and not solid matter: Jupiter, Saturn, Uranus, and Neptune.

Globular Cluster: A large and dense grouping of gravitationally bound older stars.

Heliosphere: An immense bubble within the interstellar medium that accommodates our solar system and the influences of the sun's solar wind.

Interstellar Medium: The gases and dust that permeate the vast corridors of space between and among the incalculable numbers of stars.

Light-year: The distance that light travels in one year.

Lunar Highlands (Terrae): Elevated lighter-colored regions of the moon that encompass 83 percent of its overall surface.

Lunar Maria: Dark volcanic plains that comprise approximately 17 percent of the moon's visible surface.

Magnetar: A subspecies of a neutron star believed to possess unprecedented magnetic abilities and the capacity to generate energy from its spectacular rotation.

Messier's Objects: Eighteenth-century French astronomer Charles Messier's catalog of 110 deep-sky objects, which is still widely used.

Meteoroid: A rocky or metallic space object that is smaller than an asteroid and often no bigger than a pebble.

Milky Way: The barred spiral galaxy, approximately 100,000 light-years in diameter, in which our solar system resides.

Natural Satellite (Moon): A celestial object that orbits a planet or smaller space body such as an asteroid.

Nebula: A disseminated mass of interstellar gases and dust.

Neutron Star: A star that has a solar mass four to eight times greater than our sun's, is made of neutrons, and has spectacularly collapsed under its own gravity in a supernova event.

Northern Hemisphere: The area of Earth north of the equator.

Nova: A thermonuclear explosion initiated by a white dwarf star's still-smoldering inert core when it interacts with residual hydrogen from a red giant companion.

Opposition: The positioning of a planet or another celestial body directly opposite the sun and aligned with Earth.

Perihelion: The point nearest the sun in a planet's, comet's, or another space body's orbit around it. The point farthest from the sun is known as aphelion.

Planet: A large celestial object that orbits a star.

Pulsar: A type of a neutron star, small and dense, that simultaneously maintains a potent magnetic field, furiously spins on its axis, and generates pulses of radiation.

Red Dwarf Star: A small, dim star—the most common found in the universe—that burns its fuel source slowly and lives a long time.

Red Giant Star: A temporarily enlarging and brightening old star that is nearing depletion of its life-sustaining hydrogen fuel source and is expelling its outer layers.

Reflector Telescope (Newtonian Telescope): A telescope with a wide-viewing field that employs mirrors to accumulate light and images.

Refractor Telescope: A simply designed telescope that utilizes a lens to capture light and images.

Sky Map (Star Map): A map indicating the positions of stars as seen on the celestial sphere from the perspective of Earth.

Solar System: A region in space dominated by one star's considerable mass and commensurate gravitational influence, with innumerable objects of less mass orbiting it.

Solar Wind: Fast-moving ionized particles perpetually flowing off the sun's corona—its outer atmosphere—into the ether of space.

Southern Hemisphere: The area of Earth south of the equator.

Star: A flaming orb of gases held together by gravity and typically in the midst of nuclear fusion: burning hydrogen into helium.

Sun: A second-generation yellow dwarf star, approximately 4.6 billion years old and the focus of our solar system.

Sunspots: Small, relatively cooler dark splotches that frequently appear and then disappear on the sun's surface.

Supernova: A dramatically luminous stellar event that sometimes occurs when an old and large star ceases its generation of nuclear fusion and implodes.

Terrestrial Planets: Planets in our solar system consisting of primarily rock and solid matter: Mercury, Venus, Earth, and Mars.

Variable Star: A star that from the vantage point of Earth displays varying degrees of brightness.

White Dwarf Star: A small, faint, and dense star at the end of its stellar life.

Zodiac Constellations: The thirteen constellations—areas of the sky that seemingly accommodate the sun on its annual ecliptic.

PHOTO CREDITS

Chapter 1

xii (left): © Mikhail Kokhanchikov | Dreamstime.com
xii (right): © Jose Gil | Dreamstime.com
1 (left): © Roberto Pirola | Dreamstime.com
1 (right): © Httin | Dreamstime.com
2 (left): Illustration by David Cole Wheeler
2 (right): © gary yim | Shutterstock
3 (left): Courtesy of NASA
3 (right): © Sabino Parente | Dreamstime.com
4 (left): Courtesy of NASA
4 (right): Courtesy of NASA/Johns Hopkins University Applied Physics Laboratory/
  Southwest Research Institute
5 (left): Courtesy of NASA/JPL–Caltech/University of Arizona
5 (right): Courtesy of NASA
6 (left): © Roberto Pirola | Dreamstime.com
6 (right): Courtesy of NASA/ JPL–Caltech
7 (left): Courtesy of NASA/JPL–Caltech/W. Reach (Caltech)
7 (right): © Dnally | Dreamstime.com
8 (left): Courtesy of European Space Agency & NASA
8 (right): Courtesy of NASA
9 (left): Courtesy of NASA
9 (right): © Wikipedia Commons/NASA
10 (left): Courtesy of NASA
10 (right): Courtesy of NASA/JPL–Caltech/T. Pyle (SSC)
11 (left): Illustration by David Cole Wheeler
11 (right): Courtesy of NASA GSFC

Chapter 2

12 (left): © Sandra Iacone | Dreamstime.com
12 (right): © Anthony Hall | Dreamstime.com
13 (left): © Wikipedia Commons/ ESO/Y. Beletsky
13 (right): © Changhua Ji | Dreamstime.com
14 (left): © Wikipedia Commons/Mike Peel/www.mikepeel.net
14 (right): Courtesy of NASA
15 (left): © Goofy1991 | Dreamstime.com
15 (right): © Songquan Deng | Dreamstime.com
16 (left): © Nikhil Gangavane | Dreamstime.com
16 (right): © Diego Cervo | Dreamstime.com
17 (left): © Mike Brake | Dreamstime.com
17 (right): © Axel Drosta | Dreamstime.com
18 (left): © Ktphotog | Dreamstime.com
18 (right): © Photohare | Dreamstime.com
19 (left): © Raymond Kasprzak | Dreamstime.com

19 (right): © Tatyana Chernyak | Dreamstime.com
20 (left): © Wikipedia Commons/NASA
20 (right): Courtesy of NASA
21 (left): Illustration by David Cole Wheeler
22 (left): © Sergey Peterman | Dreamstime.com
22 (right): Courtesy of Olympus Imaging America, Inc.
23 (left): © Ed Phillips | Dreamstime.com

Chapter 3

24 (left): © Wikipedia Commons/Jan Matejko–Astronomer
  Copernicus–ConversationwithGod
24 (right): © WikipediaCommons/source
  http://www.nmm.ac.uk/mag/pages/mnuExplore/PaintingDetail.cfm?ID=BHC270C
25 (left): © Wikipedia Commons/ http://commons.wikimedia.org/wiki/File:Johannes
  Kepler_1610.jpg
25 (right): © Wikipedia Commons/http://commons.wikimedia.org/wiki/File:SS–new
  ton.jpg painting by Kneller
26 (left): Illustration by David Cole Wheeler
26 (right): Illustration by David Cole Wheeler
27 (left): Illustration by David Cole Wheeler
28 (left): © Saiva | Dreamstime.com
28 (right): Illustration by David Cole Wheeler
29 (left): Illustration by David Cole Wheeler
29 (right): Courtesy of NASA, ESA, R. Windhorst (Arizona State University) and H. Yan
  (Spitzer Science Center, CalTech)
30 (left): Courtesy of NASA
31 (left): Courtesy of NASA/JPL–Caltech
31 (right): Courtesy of NASA/STScI/A. Siemigi Nowska et al. 2003
32 (left): Courtesy of NASA
32 (right): Courtesy of NASA/JPL–Caltech
33 (left): Courtesy of NASA
33 (right): Courtesy of NASA
34 (right): Illustration by David Cole Wheeler
35 (left): Courtesy of NASA
35 (right): © Evan Kirkland | Dreamstime.com

Chapter 4

36 (right): © Wikipedia Commons/ESO/Y. Beletsky/http://www.eso.org/public/
  images/moonvenuscon–potw
37 (left): © Konstantin Mironov | Shutterstock
37 (right): © Wikipedia Commons/NASA
38 (left): Illustration by David Cole Wheeler
38 (right): Courtesy of NASA

39 (right): Illustration by David Cole Wheeler
40 (left): Illustration by David Cole Wheeler
40 (right): © Wikipedia Commons/Jeff Barton
41 (left): © Photography Perspectives–Jeff Smith | Shutterstock
42 (left): © Wikipedia Commons/NASA
42 (right): Courtesy of NASA
43 (right): Courtesy of NASA/CXC/UCLA/MIT/M.Muno et al.
44 (left): Courtesy of NOAO; Infrared: NASA/JPL–Caltech/C. Engelbracht (University of Arizona)
44 (right): Courtesy of NASA
45 (right): Courtesy of NASA
46 (left): © Wikipedia/NASA/JPL–Caltech
46 (right): Illustration by David Cole Wheeler
47 (left): Illustration by David Cole Wheeler
47 (right): Illustration by David Cole Wheeler

Chapter 5
48 (left): Illustration by David Cole Wheeler
48 (right): Illustration by David Cole Wheeler
49 (left): © William Perry | Dreamstime.com
49 (right): Illustration by David Cole Wheeler
50 (right): Illustration by David Cole Wheeler
51 (left): Illustration by David Cole Wheeler
52 (left): Courtesy of Atlas Image obtained as part of the Two Micron All Sky Survey (2MASS) funded by the National Aeronautics and Space Administration and the National Science Foundation
52 (right): Courtesy of Allan Sandage, Carnegie Institution
53 (left): Courtesy of NASA/CXC/Penn State/E. Feigelson & K. Getman et al.
53 (right): Courtesy of NASA and A. Fujii
54 (left): Courtesy of Bob and Bill Twardy/Adam Block/NOAO/AURA [http://www.aura–astronomy.org]/NSF
55 (left): Courtesy of NASA/CXC/CfA/R. Hicko x et al.; Moon: NASA/JPL
55 (right): Courtesy of NASA
56 (left): Courtesy of NASA/JPL–Caltech/GSFC/SDSS
56 (right): Courtesy of NASA/ESA/G. Bacon
57 (left): © a. v. ley | Shutterstock
57 (right): Courtesy of NASA/Hubble/A. Fujii
58 (right): ©Wikipedia Commons/NASA, ESA, the Hubble Heritage (STScI/AURA)–ESA/HubbleCollaboration and A. Evans (University of Virginia, Charlottesville/NRAO/Stony Brook University)
59 (left): Illustration by David Cole Wheeler
59 (right): Illustration by David Cole Wheeler

Chapter 6
60 (left): Illustration by David Cole Wheeler
60 (right): Illustration by David Cole Wheeler
61 (left): © Damian Gil | Shutterstock
62 (left): Courtesy of NASA
63 (left): Illustration by David Cole Wheeler
63 (right): © Peresanz | Dreamstime.com
64 (left): Courtesy of ESA/NASA/and L. Calcada
64 (right): © Wikipedia Commons/Colin Jay http://www.flickr.com/photos/8725688@N04/3560612975/ Creative Commons Attribution–Share Alike 2.0
65 (left): Courtesy of NASA and Donald R. Pettit
65 (right): © Wikipedia Commons
66 (left): © David Herraez | Dreamstime.com
66 (right): © Wikipedia Commons/NASA, ESA, the Hubble Heritage (STScI/AURA)–ESA/Hubble Collaboration and A. Evans (University of Virginia, Charlottesville/NRAO/Stony Brook/original source: htt://hubblesite.org/newscenter/archive/releases/2008/16/image/at/Univers)
67 (left): © Wikipedia Commons/ Credit: Konrad Denner/Wolfgang Voges/NASA
68 (left): Courtesy of NASA
68 (right): Courtesy of ESA/NASA/and A. Nota
69 (left): © Wikipedia Commons/NASA and ESA
69 (right): Courtesy of NASA
70 (left): Illustration by David Cole Wheeler
70 (right): ©Wikipedia Commons/ESO–http://www.eso.org/public/outreach/press–rel/pr–2008/phot–44–08.html
71 (left): ©Wikipedia Commons/Roberto Mura/ http://commons.wikimedia.org/wiki/User:Roberto_Mura
71 (right): Courtesy of NASA/JPL–Caltech, D. Figer (Space Telescope Science Institute/Rochester Institute of Technology) and the GLIMPSE Legacy team of E. Churchwell, B. Babler, M. Meade, and B. Whitney (University of Wisconsin), and R. Indebetoux (University of Virginia)

Chapter 7
72 (left): © Marleen Smets | Dreamstime.com
72 (right): © Vlad | Dreamstime.com
73 (left): Courtesy of NASA
73 (right): Courtesy of NASA/Goddard Space Flight Center Scientific Visualization Studio
74 (left): © Dan Collier | Dreamstime.com
74 (right): © Ken Hurst | Dreamstime.com
75 (left): © Gleggy | Dreamstime.com
75 (right): © Terrance Emerson | Dreamstime.com
76 (left): © Bgaf72 | Dreamstime.com

118 (left): Courtesy of Francesco Parasce, ESA [http://spacetelescope.org]/STScI [http:///www.stsci.edu] and NASA

119 (left): Courtesy of NASA

119 (right): Courtesy of X–ray: NASA/CXC/Penn State/S.Park et al.; Optical: Pal.Obs. DSS

Chapter 11

120 (left): © Hamed Iravanchi | Dreamstime.com

120 (right): © Liliya Drifan | Dreamstime.com

121 (left): © Ctpaul | Dreamstime.com

121 (right): © Wyls | Dreamstime.com

122 (left): © Wikipedia Commons/author Urhixidur

122 (right): Courtesy of NASA/Goddard Space Flight Center Scientific Visualization Studio. Source data courtesy of HAO and NSO PSPT project team. HAO is a division of the National Center for Atmospheric Research which is supported by the National Science Foundation. Special thanks to Vanessa George (University of Colorado/LASP) and Randy Meisner (Michigan State University)

123 (left): Courtesy of NASA

123 (right): Illustration by David Cole Wheeler

124 (left): Illustration by David Cole Wheeler

124 (right): Courtesy of NASA Goddard Space Flight Center (NASA–GSFC)

125 (left): Courtesy of NASA Johnson Space Center (NASA–JSC)

125 (right): Courtesy of Hinode JAXA/NASA

126 (left): Courtesy of NASA

126 (right): Courtesy of SOHO, the EIT Consortium, and the MDI Team

127 (left): Courtesy of SOHO, the EIT Consortium, and the MDI Team

127 (right): Courtesy of NASA

128 (left): Illustration by David Cole Wheeler

128 (right): Courtesy of NASA

129 (left): Courtesy of NASA

129 (right): Courtesy of NASA Edwin E. Buzz Aldrin

130 (left): Courtesy of NASA

130 (right): © Elyrae | Dreamstime.com

131 (left): © Peter Dolinsky | Dreamstime.com

131 (right): © Natalia Pavlova | Dreamstime.com

Chapter 12

132 (right): © Wikipedia Commons/NASA, Hubble European Space Agency, and Akira Fujii

133 (left): © Wikipedia Commons/NASA, Hubble European Space Agency, and Akira Fujii

133 (right): © Jasenka Lukša | Shutterstock

134 (right): Illustration by David Cole Wheeler

135 (left): Illustration by David Cole Wheeler

135 (right): Illustration by David Cole Wheeler

136 (left): Illustration by David Cole Wheeler

136 (right): Illustration by David Cole Wheeler

137 (left): Illustration by David Cole Wheeler

137 (right): Illustration by David Cole Wheeler

138 (left): Illustration by David Cole Wheeler

138 (right): Illustration by David Cole Wheeler

139 (left): Illustration by David Cole Wheeler

139 (right): Illustration by David Cole Wheeler

140 (left): Illustration by David Cole Wheeler

140 (right): Illustration by David Cole Wheeler

141 (left): Illustration by David Cole Wheeler

141 (right): Illustration by David Cole Wheeler

142 (left): Illustration by David Cole Wheeler

142 (right): Illustration by David Cole Wheeler

143 (left): Illustration by David Cole Wheeler

143 (right): Illustration by David Cole Wheeler

Chapter 13

144 (left): Illustration by David Cole Wheeler

144 (right): Illustration by David Cole Wheeler

145 (left): © Jasenka Lukša | Shutterstock

146 (left): Illustration by David Cole Wheeler

146 (right): © Jasenka Lukša | Shutterstock

147 (left): Illustration by David Cole Wheeler

147 (right): © Jasenka Lukša | Shutterstock

148 (left): Illustration by David Cole Wheeler

149 (left): Illustration by David Cole Wheeler

149 (right): © Jasenka Lukša | Shutterstock

150 (left): Illustration by David Cole Wheeler

150 (right): © Jasenka Lukša | Shutterstock

151 (left): Illustration by David Cole Wheeler

151 (right): © Jasenka Lukša | Shutterstock

152 (right): Illustration by David Cole Wheeler

153 (left): Illustration by David Cole Wheeler

153 (right): © Jasenka Lukša | Shutterstock

154 (left): Illustration by David Cole Wheeler

154 (right): Illustration by David Cole Wheeler

155 (left): © Jasenka Lukša | Shutterstock

155 (right): Illustration by David Cole Wheeler

PHOTO CREDITS

Chapter 17

Chapter 18

Chapter 19

INDEX